Springer Finance

Springer

London
Berlin
Heidelberg
New York
Barcelona
Budapest
Hong Kong
Milan
Paris
Santa Clara
Singapore
Tokyo

Springer Finance

Risk-Neutral Valuation: Pricing and Hedging of Financial Derivatives
N.H.Bingham and Rüdiger Kiesel
ISBN 1-85233-001-5 (1998)

Mathematics of Financial Markets
Robert J. Elliott and P. Ekkehard Kopp
ISBN 0-387-98553-0 (1998)

Mathematical Models of Financial Derivatives
Y.-K. Kwok
ISBN 981-3083-25-5 (1998)

Guido Deboeck and Teuvo Kohonen (Eds)

Visual Explorations in Finance

with Self-Organizing Maps

With 129 Figures
including 12 Color Plates

 Springer

Guido Deboeck, PhD
3850 North River Street, Arlington, VA 22207, USA

Teuvo Kohonen, PhD
Helsinki University of Technology, Neural Networks Research Centre,
P.O. Box 2200, FIN-02015 HUT, Finland

ISBN 3-540-76266-3 Springer-Verlag Berlin Heidelberg New York

British Library Cataloguing in Publication Data
A catalogue record for this book is available from the British Library
Visual explorations in finance: with self-organizing maps.
  – (Springer finance; 3674)
  1. Finance 2. Finance – Computer simulation
  I. Deboeck, Guido J. II. Kohonen, Teuvo
  332
ISBN 3540762663

Library of Congress Cataloging-in-Publication Data
A catalog record for this book is available from the Library of Congress
Deboeck, Guido.
  Visual explorations in finance with self-organizing maps/Guido J. Deboeck and Teuvo
Kohonen (eds.).
       p. cm. – (Springer finance)
  Includes bibliographical references and index.
  ISBN 3-540-76266-3 (case bound)
  1. Finance – Decision making – Data processing. 2. Neural networks
(Computer science) 3. Self-organizing systems. I. Kohonen, Teuvo.
II. Title. III. Series.   98-11388
HG4012.5.D43 1998
332'.0285 – dc21 CIP

© Springer-Verlag London Limited 1998
Printed in Great Britain
2nd printing 2000

Typeset by The Midlands Book Typesetting Company, Loughborough
Printed and bound at the Athenæum Press Ltd., Gateshead, Tyne & Wear
12/3830-54321 Printed on acid-free paper SPIN 10759253

Dedication

Rose-Maelle F. (Haiti)
Tafadzwa G. (Zimbabwe)
Kohinoor A. (Bangladesh)
Alenjie A. (Philippines)
Antony V. (Peru)
Judy Njoki M. (Kenya)
Murtopo (Indonesia)

Any royalties we receive from this book will be applied to the health and education of our foster kids through Childreach Inc., the US member of PLAN International. Childreach Inc., a global, child-focused development organization with counterparts in 12 countries contributes to programs that help over one million children in 40 countries in the world. To become a sponsor or learn more about Childreach visit *www.childreach.org.*

Foreword

Professor Robert Hecht-Nielsen at WEBSOM'97, Espoo, Finland Photo: GJD, Helsinki, June 1997

Often, when a technology becomes commercially valuable it ceases to be an academic subject. For such subjects, all leading-edge knowledge and research resides behind closed company doors. Current examples of such technologies include microprocessors, rocket engines, and automobiles. State-of-the-art academic research in such fields simply does not exist (although advanced research on far future concepts is alive and well).

This book describes a rich selection of financial, economic and marketing applications of Teuvo Kohonen's Self-Organizing Map (SOM). In these application areas, SOMs have proven to be of significant economic value. As a result, this area of work is rapidly becoming a non-academic technology.

This book is perhaps the last look we will have at near-state-of-the-art SOM applications in these areas. It is thus apt that such a timely and valuable collection of work be edited by Guido Deboeck, a leading exponent of the use of computational intelligence methods in finance and economics, and by the originator of SOM, Teuvo Kohonen. Their efforts and that of their contributing authors will pay off handsomely for us all.

San Diego, California   Robert Hecht-Nielsen
September 1997

Preface by Guido Deboeck

At the 1990 IJCNN (International Joint Conference on Neural Networks) in Paris I met Teuvo Kohonen for the first time. I heard him speak very enthusiastically about an algorithm that could organize data based on a computer algorithm called self-organizing maps (SOM). At the 1993 IJCNN in Seattle I suggested to Teuvo that self-organizing maps might be useful to apply to economic data, in particular the World Development indicators, published annually by the World Bank. Following our conversation I sent the World Development indicators to Helsinki. Two years later I found a poverty map constructed on the basis of the World Development indicators in the first edition of Teuvo's *Self-Organizing Maps*. I was very pleased to see this first poverty map based on SOM.

In October 1996 Teuvo and I met in Arlington. On a nice fall weekend in Virginia we discussed applications of self-organizing maps to finance and economics. Over 3000 papers have been written about SOM but most are in engineering; few focus on applications in finance or economics.

In January 1997 I spent 48 hours in Helsinki: it was cold and dark, the streets were covered with ice, getting around was dangerous, except for the Finns who have gotten used to it. The meeting with Teuvo was warm; we put together an initial outline and enjoyed a very nice dinner in a Lapp restaurant. In February we invited several people to contribute chapters. Most of the papers in this book were written in April and June. In early June I was back in Helsinki and met the entire Neural Network Research team at the WEBSOM'97 conference. This was a very productive meeting, and this time I did get to see Helsinki! The outline for this book changed several times: many chapters were added including some on using SOM for real estate investments and marketing. At the WEBSOM conference I also met Dr Gerhard Kranner and Johannes Sixt of Eudaptics GmbH who provided me with access to Viscovery SOMine. Thanks to it I was able to analyze the Chinese consumer data that Bernd Schmitt provided in June 1997.

From the initial idea in October 1996 to a complete manuscript by September 1997, this work involved collaboration between about 20 people from Spain to Finland with the editors in Helsinki and Arlington. This book is the product of *virtual teamwork*, i.e. a team with self-created objectives, motivation and discipline, without structure or management. While this concept could be used in other contexts the full meaning of "virtual teamwork" is yet to be understood by many.

After my first book (*Trading at the Edge*) was published in April 1994, I was invited to make presentations in New York, London, Frankfurt, Geneva, Singapore, Seoul, Nagoya and Beijing. On these trips the most rewarding encounters were with people who wanted to apply neural networks in their areas of interest, write a PhD, or a book on the topics explored in *Trading at the Edge*. I hope the same will happen with this one. May many take up the challenge of improving on these novel ways of looking at financial, economic or marketing data because *knowledge is power, but to effectively acquire it, use it and expand it, new modes of learning, synthesizing and collaboration will need to be devised.*

Arlington, Virginia   Guido Deboeck
December 1997

Preface by Teuvo Kohonen

Professor Teuvo Kohonen at WEBSOM'97, Espoo, Finland
Photo: GJD, June 1997

Prior to 1993 I had only heard of Guido Deboeck and his application of neural networks to financial problems. Therefore, I was very much taken when he approached me at the IJCNN'93 in Seattle and asked me to do more work in financial applications. Actually that was the right time for us, since we were reorganizing our laboratory. Besides the World Bank data, we were starting bankruptcy analyses on small and medium-sized enterprises with the Finnish Foundation, Kera Ltd.

In 1996 Guido and I established close cooperation that still continues. Actually I engaged several bright and enthusiastic junior collaborators to this project – Samuel Kaski, Antti Vainonen, Janne Nikkila, the SOM Toolbox developers (Esa Alhoniemi, Johan Himberg, Kimmo Kiviluoto, Jukka Parviainen and Juha Vesanto), and the WEBSOM group (Samuel Kaski, Timo Honkela and Krista Lagus). Dr Samuel Kaski spent a lot of time collaborating with Guido through e-mail and over the Internet. This was quite a nice experience; we were working closely together between Arlington (Virginia) and Espoo (Finland); transferring outlines, data, ideas and revisions, while we did not see each other for months. Isn't this an information age!

My personal share of the editing work has been much less than that of Guido. I have been in the lucky situation that the people around me have taken so many responsibilities, and I am very much obliged to all of them.

Espoo, Finland
December 1997

Teuvo Kohonen

Acknowledgements

Teuvo Kohonen and Samuel Kaski contributed to my initial understanding of self-organizing maps. Without their support and efforts this book would never have been written. Carlos Serrano-Cinca's work on financial and economic applications inspired me to undertake this project.

Many people have contributed to this book. I wish to thank in particular Eric de Bodt in Belgium; Marina Resta and Aristide Varfis in Italy; Carlos Serrano-Cinca in Spain; Serge Shumsky in Russia, Eero Carlson, Timo Honkela, Samuel Kaski, Kimmo Kiviluoto, and Anna Tulkki in Finland; and Bernd Schmitt, who granted me access to the Chinese consumer data in Shanghai. All of them have made pioneering contributions. Their work has encouraged me to put these applications in one volume. Several others have contributed to chapters in this book (Marie Cottrell, Philippe Grégoire, Jari Kangas, Jorma Laaksonen, Krista Lagus, Kari Sipilâ, Cristina Versino, A.V. Yarovoy) or have reviewed earlier drafts of the manuscript (Masud Cader, Luc De Wulf, Mark Embrechts, Rik Ghesquiere, Gerhard Kranner, Jack L. Upper, J.D. Von Pischke). All deserve credit for contributing to this book.

Many thanks are also due to the publisher, editor and production coordinator. Dr Susan Hezlet of Springer-Verlag made the publishing process really easy; Lyn Imeson performed a superb job in editing the manuscript; Nick Wilson managed the print process and was very helpful in improving the quality of all the figures; and Vicki Swallow coordinated all communications between the editors, the publisher, and everyone else involved in this process. Needless to say that all remaining errors or misrepresentations are mine.

A word of thanks also to my former colleagues at the World Bank: Sven Sandstrom, Ian Scott and Hywel Davies have each in their own way contributed to making this project feasible. Luc De Wulf, Adrien Goorman, Rik Ghesquiere, Paul Staes and many others have through their friendship and moral support continued to provide invaluable help over many years.

This project could not have come to fruition without the support from my family. Toni, my oldest son, read and edited all the original draft contributions; his degree in aeronautical engineering and prior work in neural networks, genetic algorithms, and fuzzy logic was invaluable. Pascal read the Introduction and has made several down-to-earth suggestions to shorten the text; he also helped with artwork, and the wiring of my home office. My daughter, Nina, who likes playing music and writing stories, would probably have preferred a fiction book about horses. My wife, Hennie, who has a lot of patience, helped us to survive this project by letting me, right in the middle of it, take the kids on a camping trip around the world. Our friend, Charing, took care of all the flowers, the birds and the fish, and kept the candles lit while I continued the writing and surfing the web. Thank you.

Arlington, Virginia   Guido Deboeck
April 1998

Contents

Part 1: Applications

Contributing Authors

Eero Carlson
National Land Survey of Finland
Eero Carlson obtained an MSc in Engineering at the Helsinki University of Technology in Finland in 1970. Since 1973 he has worked on the development of a Finnish Geographic Information System (GIS) and since 1990 on neural networks. His main interests are real estate appraisal with Self-Organizing Maps and the integration of SOM and GIS. Mr. Carlson is the author of about 15 publications on neural networks.
e-mail: carlson@atk.nls.fi

Eric de Bodt
Professor of Finance at Université Catholique de Louvain in Belgium Eric de Bodt received his PhD in Applied Economics in 1992. His doctoral thesis, entitled "Le traitement de l'information marketing et financière dans la décision d'investissement. Une intégration par les systèmes experts" was sponsored by a leading chemical company. In 1997, he was ranked first in the Agrégation Nationale Française in Management Science. During the last few years he has worked with the SAMOS Research Center at the University of Paris. He has concentrated on the application of neural networks in finance and is the author of several articles. His field of interest includes the application of information technologies in finance. Eric de Bodt is a member of the editorial board of the Journal of Computational Intelligence in Finance, a reviewer for Neural Processing Letters, and a member of several scientific committees of International Conferences.
e-mail: debodt@fin.ucl.ac.be

Guido Deboeck
Author, multimedia producer, consultant
Guido Deboeck is an expert on advanced technology and its applications for financial engineering and management. For 20 years he has been an innovator and advisor on technology to the World Bank

in Washington. He has taught numerous courses for senior executives and has published several articles and a book entitled *Trading on the Edge: Neural, Genetic and Fuzzy Systems for Chaotic Financial Markets.* He is also a multimedia producer who has created several CD-ROMs, including the *Trading Navigator,* and *Business Innovation,* a multimedia application developed for the World Bank. His experience with the management of complex environments includes the management of technology of the Investment Department, renovation of a trading floor, and development of trading systems for the World Bank. He also contributed to the design of the trading floors in the Asian Development Bank and the People's Bank of China. Prior to this, he spent over 10 years working on information systems for planning, monitoring and evaluation of poverty projects around the world. Guido Deboeck is a regular speaker at international conferences. He has a PhD in Economics from Clark University (Mass) and a graduate degree from the Catholic University of Leuven in Belgium. He is interested in eco-tourism, photography, and home beer brewing.
e-mail: gdeboeck@worldbank.org

Timo Honkela
Research Scientist, Neural Networks Research Centre, Helsinki University of Technology
Timo Honkela obtained his MSc degree in Information Processing Science at University of Oulu, Finland in 1989. His main interests and research experience are natural language processing. From 1987 to 1989 he was responsible for the design of the semantic processing component in a major project developing a natural language database interface for Finnish. From 1990 to 1994 he was a research scientist at the Technical Research Centre of Finland. At that time he was a project manager in the Glossasoft project in Linguistic Research and Engineering, a program funded by EU commission. Since 1995 Honkela has been with the Neural Networks Research Centre of the Helsinki University of Technology where he is doing research related to various aspects of natural language interpretation using Self-Organizing Maps. Honkela is also the chairman of the Finnish Artificial Intelligence Society.
e-mail: timo.honkela@hut.fi

Samuel Kaski
Research Associate, Neural Networks Research Centre, Helsinki University of Technology
Samuel Kaski received his degree of Doctor of Technology in Computer Science at Helsinki University of Technology in 1997. His main research interests are neural networks, especially self-organizing maps, and their applications in statistics and data mining. While working at the Neural Networks Research Centre he has participated in a variety of research

projects including speech recognition, theoretical physiological modeling, analysis of electromagnetic signals of the brain, analysis and visualization of macroeconomic statistics, and automatic organization of textual document collections. The common theme in most of his projects has been exploratory data analysis and visualization of structures in the data using self-organizing maps. Dr Kaski has published several research papers connected to both the theory and applications of the methodology.
e-mail: sami@guillotin.hut.fi

Kimmo Kiviluoto
Researcher, Laboratory of Computer and Information Sciences, Helsinki University of Technology
Kimmo Kiviluoto received his MSc degree from the Helsinki University of Technology in 1996 and is now pursuing his PhD degree at the same university. His research interests include neural networks, unsupervised learning and self-organization, and their applications in financial engineering.
e-mail: kkluoto@nucleus.hut.fi

Teuvo Kohonen
Professor of Computer Science, and Professor at the Academy of Science of Finland
Professor Kohonen's research areas are associative memories, neural networks, and pattern recognition, in which he has published over 200 research papers and four monographs. Professor Kohonen has introduced several concepts in neural computing including fundamental theories of distributed associative memory, optimal associative mappings, the learning subspace method, self-organizing feature maps, learning vector quantization, algorithms for symbol processing, adaptive-subspace SOM, and SOMs for exploratory textual data mining. His best known application is a neural speech recognition system. He is the recipient of several honorary prizes. He received a Honorary Doctorate degree from the University of York, UK as well as from Åbo Akademi, Finland. He is a member of Academia Scientiarum et Artium Europaea, a titular member of the Academié Européenne des Sciences, des Arts et des Lettres, a member of the Finnish Academy of Sciences and the Finnish Academy of Engineering Sciences, an IEEE Fellow, and a Honorary Member of the Pattern Recognition Society of Finland, as well as the Finnish Society for Medical Physics and Medical Engineering. He was the First Vice-President of the International Association for Pattern Recognition (1982–84), and acted as the first President of the European Neural Network Society from 1991 to 1992. Professor Kohonen's newest book on *Self-Organizing Maps* was published in the Springer Series on Information Sciences in 1995.
e-mail: teuvo.kohonen@hut.fi

Marina Resta
Institute of Financial Mathematics, University of Genoa, Italy
Maria Resta received a MS degree in Economics from the University of Genoa, Italy in 1995. She is currently with the Institute of Financial Mathematics, Faculty of Economics at the University of Genoa. Her research interests include neural networks, nonlinear dynamics and chaotic systems. She is currently working on trading solutions for financial markets forecasting via artificial intelligence tools.
e-mail: resta@economia.unige.it

Bernd Schmitt
Associate Professor in Marketing, Columbia Business School, New York, and BAT Chair of Marketing at CEIBS, Shanghai
Bernd Schmitt is an expert on international marketing, corporate identity, advertising and consumer behavior. He has taught numerous executive programs and seminars for international corporations in the US, Europe and Asia. He has been a visiting professor at MIT, in Germany, Poland and Hong Kong. In the People's Republic of China, he has taught hundreds of executives over a period of five years. Dr Schmitt has contributed more than 30 articles to management journals. His research on corporate identity, branding and international marketing has been covered in *The Economist*, the *New York Times*, the *LA Times*, the *Washington Post* and the *South China Morning Post*, as well as on US, Russian and Chinese TV.
e-mail: bschmitt@research.gsb.columbia.edu

Carlos Serrano-Cinca
Lecturer, University of Zaragoza, Spain
Carlos Serrano-Cinca is a Lecturer in Accounting and Finance at the University of Zaragoza in Spain. He is also a Visiting Lecturer at the Department of Accounting and Management Science at the University of Southampton, UK. He received his PhD in Economics and Business Administration from the University of Zaragoza in 1994. His Doctoral Thesis, entitled "Neural Networks in Financial Statement Analysis" received the prize for the outstanding thesis of the year. His research interests include applications of artificial intelligence in accounting and finance and analysis of quantitative financial information with multivariate mathematical models. Dr Serrano-Cinca has published several articles in *Decision Support Systems, Neural Computing Applications*, the *European Journal of Finance* and *Omega, the International Journal of Management Science*. He has also published and served as an ad hoc reviewer for academic journals in accounting and finance.
e-mail: serrano@posta.unizar.es

Serge Shumsky

Group Leader of the Neural Network Group at Lebedev Physics Institute, Russia

Serge A Shumsky develops applications of neural networks in finance, telecommunications and technology. Shumsky obtained a PhD in Plasma Physics from Lebedev Physics Institute in 1988. He joined the Lebedev Physics Institute in 1981 after receiving an Honors degree from Moscow Physical Engineering Institute in Nuclear Physics. In 1991 he switched to the Theory of Computations and Neural Networks. Since 1994 he has been a Project Manager of a project on the "Development of a High Capacity Content Addressed Neural Network Memory" of the International Science and Technology Center. His current fields of interests are complex systems, theory of computations, and theory and applications of neural networks.

e-mail: shumsky@neur.lpi.msk.su

Anna Tulkki

ADP Analyst, COMPTEL, Finland

Anna Tulkki obtained an MSc in surveying at Helsinki University of Technology in 1996. Since then she has worked as a teacher of surveying in a vocational school and is now a trainer/instructor at Comptel, which is a software company specializing in software development for telecommunications applications.

e-mail: anna.tulkki@comptel.fi

Introduction

Knowledge Discovery

Collaboration in cyberspace and commerce on the worldwide web are expanding. As of December 1997 a large survey conducted by Nielsen Media Research indicates that about 58 million adults in the U.S. and Canada are using the Internet. This high estimate suggests tremendous growth in the use of the worldwide web. Nielsen's biannual study in cooperation with CommerceNet, an electronic-commerce trade group, is based on interviews with some 9000 users. It indicates a roughly 15% increase in users compared with six months earlier. This finding implies an annual growth rate in the 32% range or an approximate annual doubling of users, at least over the past few years.

This rapid increase in usage of the worldwide web provides expanded opportunities for information dissemination. The number of pages on the worldwide web has grown from a few thousand to millions over the past three years. But maybe *pages* is no longer a proper way of thinking or measuring the scope of the information on the worldwide web.

Commerce on the web is also expanding. There are an increasing number of cybershops and services offered over the web. Popular uses of the worldwide web are travel planning (making flight reservations, renting a car, booking a hotel room); electronic banking and trading; and shopping (e.g. for hardware, software, books, wine). For example, *Virtual Vineyard*, at *www.virtualvin.com*, is a website that contains a nice selection of wines, allows ordering on-line, and shipments are delivered to your door. The largest electronic bookstore on the web, at *www.amazon.com*, claims (at the time of this writing) over two million books or fifteen times more titles than can be found in the largest bookstores in the world. There are many examples of increased conveniences, time and money saving (not to mention the tax savings) through the worldwide web.

A more important implication of the growing use of the web is the increased potential for collaboration, teamwork, and creation of new enterprises. An example of this may be our own experience. Four years ago *Trading at the Edge* (Deboeck, 1994a) was edited based on a dozen contributions from authors around the globe. Contributions were submitted on floppy disks, by snail mail or facsimile. This new book involved collaboration between two editors on two different continents; required collaboration between twenty contributors located from Finland to Spain; and was based on electronic file exchanges, remote processing and analysis of data. For example, data collected in April/ May 1997, in the streets of Beijing and Shanghai, was transferred electronically to Arlington. This data was analyzed using programs in Finland; initial results were reviewed in Vienna, editing was done in Arlington and the results submitted electronically to Columbia University in New York. All of this was accomplished in weeks after the survey and the data was entered in Shanghai.

Another example was the collaborative editing and writing that took place over the summer in 1997: comments from reviewers in Singapore were transferred to the contributors in Europe; these comments would often trigger rewriting of sections, which were then resubmitted for the draft manuscript that was assembled in Arlington. It is amazing how small the world becomes if one can have *various windows* into servers around the world!

In this collaborative production and writing effort, teamwork emerged among people who never met nor even knew each other. The common thread among them was an interest in a topic, an approach or method, in this case for better visualization and reduction of high-dimensional financial data, experience in applying this method, and a willingness to share knowledge. This common interest did not exist before the editors dreamed it up on a fall weekend in Virginia.

This new paradigm of collaborative work, sometimes called *virtual teamwork*, may be a positive outcome from the drive towards cost cutting, downsizing and re-engineering of organizations in the earlier part of the 1990s. In 1996, John Micklethwait and Adrian Wooldridge wrote that these movements towards cost cutting, downsizing and re-engineering were pushed in many corporations by management gurus, most of whom were "witch doctors" or "charlatans" (Micklethwait and Wooldridge, 1996). Like witch doctors, the management gurus of the early 1990s predicted a future of chaos and uncertainty. Their own work made this even more likely. Indeed, many of these management gurus were ill equipped to use the technology and communication capabilities that were so rapidly evolving via the worldwide web. Several had no idea about the wealth of information and tools that were available to make organizations more effective. The anxieties they imposed on others were often personal anxieties or discomfort with the directions the technology and the worldwide web was taking. It is doubtful that the management gurus of the 1990s, whether true management gurus, witch doctors or charlatans, will have any substantial impact on this new paradigm of virtual teamwork that is evolving. It is doubtful that they will have any impact on the opportunities for collaborative work and the creation of new enterprises that are created by the worldwide web.

Thus we find at the dawn of the twenty-first century almost unlimited access to information, vast expanded possibilities for collaborative work, new enterprises, improved knowledge management, and knowledge discovery.

Knowledge management has become the centerpiece of change efforts in many organizations. Effective knowledge management requires, however, that an organization defines

- what value it intends to provide;
- to whom it plans to provide this value (in other words, it must have clear targets because no one can serve five billion customers);
- what resources and/or partnerships it can mobilize to provide the added value;
- how it actually will produce added value.

There are few organizations that meet this challenge.

In *The Knowledge-creating Company*, Ikujiro Nonaka and Hiritaka Takeuchi (1995) show how companies can add value to their bottom line by creating new knowledge organizationally. Nonaka and Takeuchi point out the difference between *explicit knowledge*, what can be found in databases, reports or manuals, and *tacit knowledge*, that is knowledge learned by employees from experience and communicated indirectly. The key to success in the future is, according to Nonaka and Takeuchi, to learn how to convert *tacit* into *explicit* knowledge. Most organizations are, however, much better at designing information systems than at converting tacit into explicit knowledge. Many knowledge management systems are doing little more than repackaging information, or what some have called *microwaving information*.

The expanded use of the worldwide web is demanding novel methods for knowledge capturing, better ways of synthesizing, and improved ways of visualizing massive amounts of information. *The focus in this book is exactly on these. This book is about automated ways to extract, synthesize, and visualize massive amounts of information. It focuses on the discovery of new structures and patterns in data; and on extracting new knowledge from existing data or experiences.*

This book does not assume any specialized knowledge of particular methods nor of any specialized quantitative or statistical techniques. On the contrary, it explains how to extract knowledge and information from existing data. It starts with a simple example on choosing the best Scotch whiskies; and then provides, in Part 1, lots of other examples of improved methods for synthesis and display of information. Only thereafter are the methods and software discussed. Part 2 also discusses the best practices in data mining.

The methods discussed in this book belong to the general class of neural network models. Neural networks are a collection of mathematical techniques that can be used for signal processing, forecasting, and clustering. Neural networks can be thought about as non-linear, multi-layered, parallel regression techniques. In simple terms, neural network modeling is like fitting a line, plane or hyperplane through a set of data points. A line, plane or hyperplane can be fitted through any set of data and define relationships that may exist between (what the user chooses to be) the *inputs* and the *outputs*; or it can be fitted for identifying a representation of the data on a smaller scale. There are two classes of neural networks: supervised and unsupervised neural nets.

Supervised neural nets are techniques for extracting input–output relationships from data. These relationships can be translated into mathematical equations that can be used for forecasting or decision-making. The user identifies the desired outputs. The net learns through an adaptive, iterative process to detect the relationship(s) between the given inputs and the outputs. Once a neural net has been trained in this fashion, it can be used on data that it has never seen or be imbedded in a program for automated decision-support.

Several examples of supervised neural networks are described in *Trading on the Edge: Neural, Genetic and Fuzzy Systems for Chaotic Financial Markets* (Deboeck, 1994). Some of the main themes in that book are:

- Neural networks are powerful tools for pattern recognition and forecasting.
- Genetic algorithms, which are natural ways to evolve, allow the extraction of rules, strategies, and/or the optimization of systems aiming at multiple objectives under one or more constraints.
- Fuzzy logic provides for automated extraction of fuzzy rules and use of fuzzy expert systems.
- These algorithms and the rapid evolution of audio-visual computing, data sonification, virtual reality, and the like continue to challenge our thinking about knowledge and information management.

Unsupervised neural networks are techniques for classifying, organizing and visualizing large data sets.

An example of an unsupervised neural network technique is Self-Organizing Map (SOM). This approach has been around since the early 1980s and has been widely applied in engineering and many other fields. Many applications of unsupervised neural networks and SOM can be found in Teuvo Kohonen's *Self-organizing Maps* (Kohonen, 1997).

Kohonen's book focuses on the methodology of unsupervised neural networks. In this book we discuss exploratory data analysis, data mining, and provide an overview of traditional methods for clustering and visualizing data. We then describe self-organizing maps, and their advantages for financial, and economic and market analyses. This book shows how SOM can be used to

- analyze the financial statements of companies;
- analyze investment opportunities;
- predict long-term interest rates and bankruptcies;
- select investments in mutual funds and stocks;
- appraise real-estate properties;
- analyze markets based on consumer preferences and attitudes;
- segment customers or clients;
- support strategic marketing and market analysis.

This book shows how structures can be discovered in economic data; how to improve risk and portfolio management. Finally, it shows how the SOM approach can be used for improving the organization of large document collections and knowledge management.

This book is about detecting new knowledge and information from data and the better display of massive amounts of data. It shows the way towards the automation of knowledge discovery, and the translation of *tacit* into *explicit* knowledge. It contains suggestions for creating *knowledge maps*. This book breaks new ground in finance, economics, and marketing applications. It brings together the experience gained by many researchers, professionals, and business executives in applying advanced analytical tools for more effective knowledge and information management.

Exploratory Data Analysis and Data Mining

Exploratory data analysis and data mining can be used for *knowledge discovery*, that is, the whole interactive process of discovery of novel patterns or structures in the data. There is often confusion about the exact meaning of the terms "exploratory data analysis", "data mining" and "knowledge discovery". At the first international conference on knowledge discovery in Montreal in 1995 it was proposed that the term "knowledge discovery" be employed to describe the whole process of extraction of knowledge from data. In this context knowledge means relationships and patterns between data elements. It was further proposed that the term "data mining" should be used exclusively for the discovery stage of the process.

A more accurate definition of knowledge discovery would thus be *the non-trivial extraction of implicit, previously unknown and potentially useful knowledge from data.* Knowledge discovery is a multi-disciplinary approach involving machine learning, statistics,

database technology, expert systems, and data visualization. All of these
may make a contribution to the extraction of new knowledge.

The core of the process in exploratory data analysis and data
mining consists of a multitude of steps starting from setting up the
goals to evaluating the results. It may also involve a feedback loop that
reformulates the goals based on the results. Depending on the goals
of the process any kind of algorithms for pattern recognition,
machine learning, or multivariate analysis can be used. The key in
data mining is the actual discovery of previously unknown structures
or patterns.

In this book we will primarily use the SOM approach for the search
of patterns in large data sets. The main reason for using SOM for
exploratory data analysis and data mining are:

- it is a numerical instead of a symbolic method;
- it is a non-parametric method;
- no a priori assumptions about the distribution of the data need to
  be made;
- it is a method that can detect unexpected structures or patterns by
  learning without supervision.

We will also demonstrate that the results from SOM can often be
improved if used in conjunction with traditional statistical techniques
or combined with more advanced techniques, i.e. supervised neural
networks, genetic algorithms, or fuzzy logic. Several chapters in this
book will discuss hybrid applications of SOM.

Before we define SOM in more detail it may be useful to provide a
brief overview, in non-mathematical terms, of some of the traditional
methods for clustering and visualization of data.

Traditional Methods

There are several methods in statistics to synthesize data sets or
statistical tables. The simplest methods produce simple summaries of
the data. For example, the smallest and largest data values, the median,
the first and third quartiles. Such simple methods are very useful for
summarizing low-dimensional data sets. If the dimensionality of the
data is large then it is more difficult to synthesize and visualize the data.
In this book we concentrate on methods that can be used for
synthesizing and visualizing large multivariate data sets, and ways to
detect and illustrate *structures* within the data.

Usually data samples consist of multiple values. The values in a data
set may correspond to a set of statistical indicators. Such a set of
indicators can be expressed as a *vector*, which simply means an ordered
set of numeric values.

A data vector represents points in an n-dimensional space. As long as there are only two or three dimensions it is quite easy to make simple two- or three-dimensional graphs. However, if the dimensionality of the data is larger then it is difficult to plot a vector or the relations between different vectors. This is precisely why other visualizing methods are needed.

In common methods for visualization, each dimension of a high-dimensional data set governs some aspect of the visualization and then integrates the results into one. These methods can be used to visualize different kinds of high-dimensional data [0.01]. The major drawback of most methods is that they do not reduce the amount of data. If the data set is large, the display consisting of all the data items will be incomprehensible. These methods can, however, be useful for illustrating some summaries of the data set.

To reduce the amount of data by categorizing or grouping similar data items together is called *data clustering*. Grouping of data via clustering is pervasive since humans also process information that way. One of the motivations for using clustering is to automate the construction of categories or taxonomies. Clustering may also be used to minimize the effects of human biases or errors in the grouping process. An overview of clustering techniques can be found in Box 1.

Other methods exist that can be used for reducing the *dimensionality* of the data vectors. Some of these methods are called *projection methods*. The goal of projection methods is to represent the input data in a lower-dimensional space in such a way that certain properties of the structure of the data are preserved as faithfully as possible. Projections can be used to visualize the data if a sufficiently small dimensionality for the output display is chosen. A brief overview of projection methods can be found in Box 2.

This brief overview of traditional methods for clustering and visualizing of large data sets is sufficient background for the introduction of self-organizing maps in the next section.

Box 1: Clustering methods in a nutshell (prepared by S. Kaski)

Clustering methods can be divided into two basic types: *hierarchical* and *non-hierarchical* clustering methods. Within each there exists a wealth of different approaches and algorithms.

• **Hierarchical clustering** proceeds successively by merging smaller clusters into larger ones, or by splitting larger clusters into smaller ones. The clustering methods differ by the rule used to decide which smaller clusters are merged or which of the larger clusters are split. The end result of the algorithm is a tree of clusters called a *dendrogram*, which shows how the clusters are related.

When two small clusters are merged, a higher level is created to the dendrogram and the representation of the merged cluster at the new level is connected to the representations of the clusters in the lower level. By cutting the dendrogram at a desired level one obtains a clustering of the data items into different groups.

• **Non-hierarchical clustering** attempts to directly decompose the data set into a set of disjoint clusters. The goal of this algorithm is to assign clusters to the densest regions in the data space, i.e. to define a cluster where there is a large amount of similar data items. Another possible approach involves minimizing some measure of dissimilarity of the samples within each cluster while maximizing the dissimilarity of different clusters [0.02].

K-means is one type of non-hierarchical clustering. To approximate the density of high-dimensional input vectors using a smaller number of suitably selected *reference, model,* or *code book vectors* one minimizes the average *quantization error*. If the difference between the input sample vectors and the reference vectors is then defined as an error, then by comparing all input vectors with all reference vectors, one can identify the model vectors for which the *distance* is the smallest. This vector can be called the *winner*. In practice, each cluster can be represented by one or several reference vectors.

A problem with *K*-means clustering is that the choice of the number of clusters is critical: quite different kinds of clusters may emerge when *K* is changed. Good initialization of the model vectors is crucial (some clusters may even be left empty if the initial values lie far from the distribution of data). Another problem with *K*-means clustering (and clustering methods in general) is that the interpretation of the clusters may be difficult. Most clustering algorithms prefer certain cluster shapes, and the algorithms then tend to assign the data to clusters of such shapes even if there were no clusters in the data. If the goal is not just to compress the data but also to make inferences about its structure, it is essential to analyze whether the data exhibits a clustering tendency. Furthermore, the results of the cluster analysis need to be validated.

Box 2: Projection methods in a nutshell (prepared by S. Kaski)

There are two types of projection methods: linear and non-linear methods.

• **Linear projection methods**

If an *n*-dimensional data set is represented as an *n*-dimensional space, a subspace of this *n*-dimensional space is a two-dimensional (i.e. a plane) or one-dimensional (i.e. a line) space. A data set can be represented in a subset of vectors that constitute a linear subspace of lower dimensionality. Each vector in an m-dimensional linear subspace (where m is less than n) is a linear combination of m independently selected basis vectors. One method for displaying high-dimensional data vectors as a linear projection onto a smaller-dimensional subspace is principal component analysis.

In *Principal Component Analysis* (PCA) each component of the projected vector is a linear combination of the components of the original data item; the projection is formed by multiplying each component by a certain fixed scalar coefficient and adding the results together. Mathematical methods exist for finding the optimal coefficients such that the variance of the data after the projection will be preserved, whereby it is also closest to the variance of the original data. PCA is a standard method in data analysis; it is well understood, and an effective algorithm for computing the projection. Most textbooks on statistics contain a section on PCA, and most of the general statistical computer programs contain routines for computing it. Even *neural* algorithms for PCA have been devised [0.03].

• Non-linear projection methods
If the data set is high-dimensional and its distribution highly unsymmetrical, it may be difficult to visualize the structures of its distribution using linear projections onto a low-dimensional display. Several approaches exist for reproducing non-linear, high-dimensional structures of data on a low-dimensional display. The most common methods map each data item as a point in the lower-dimensional space, and then try to optimize the mapping so that the distances between the image points would be as similar as possible to the original distances of the corresponding data items. Various methods only differ in how the different distances are weighted and how the representations are optimized.

Multi-dimensional scaling (MDS) is one example of a non-linear projection method. MDS refers to a group of methods that is widely used in behavioral, econometric, and social sciences to analyze subjective evaluations of pairwise similarities of entities. The starting point of MDS is a matrix consisting of the pairwise dissimilarities of the entities. Here only distances between data items that have been expressed are considered. However, in MDS the dissimilarities need not be distances in the mathematical sense. MDS is perhaps most often used for creating a space where the entities can be represented as vectors when only some evaluations of the dissimilarities of the entities are available. In visual clustering the goal is not merely to create a space which would represent the relations of the data faithfully, but also to reduce the dimensionality of the data to a sufficiently small value to allow visual inspection of the set. MDS can be used to fulfill this goal [0.04].

Another non-linear projection method, closely related to MDS, is *Sammon's mapping*. This method also tries to match the pairwise distances of the lower-dimensional representations of the data items to their original distances. The difference between Sammon's mapping and the metric MDS is that in Sammon's mapping the preservation of small distances is emphasized. In the metric MDS method larger distances are given relatively more weight when the projection is computed, whereas in Sammon's mapping the contributions are normalized [0.05].

Self-Organizing Maps

A self-organizing map (SOM) is a feedforward neural network that uses an unsupervised training algorithm, and through a process called self-organization, configures the output units into a topological representation of the original data. SOM belongs to a general class of neural network methods, which are non-linear regression techniques that can be trained to learn or find relationships between inputs and outputs or to organize data so as to disclose so far unknown patterns or structures.

SOM is a neural network technique that learns without supervision. Supervised neural network techniques require that one or more outputs are specified in conjunction with one or more inputs to find patterns or relations between data. In contrast, SOM reduces multi-dimensional data to a lower-dimensional map or grid of neurons.

The SOM algorithm is based on unsupervised, *competitive learning*. It provides a topology-preserving mapping from the high-dimensional space to map units. Map units, or neurons, usually form a two-dimensional grid and thus the mapping is a mapping from a high-dimensional space onto a plane. The property of *topology preserving* means that a SOM groups similar input data vectors on neurons: points that are near each other in the input space are mapped to nearby map units in the SOM. The SOM can thus serve as a clustering tool as well as a tool for visualizing high-dimensional data.

Kohonen has suggested on several occasions that a self-organizing map can be thought of as a *display panel*, with lamps for each neuron in the output layer. For example, in a lecture he gave on SOM at IJCNN in Paris (1991), he said:

> The SOM algorithm analyzes the input data and displays this information on the panel, as if the corresponding lamps were lit. An example is the visualization of speech, which as a matter of fact, was the first real application of SOM. Imagine that you control the "lamps" of a display panel that reflects natural speech. If you capture with a microphone acoustic spectra of speech, like Professor Higgins did with old mechanical means in the musical *My Fair Lady*, then for each sound that is captured you can compute a characteristic spectrum that looks like a squiggle. For

those sounds that have predominantly low tones the squiggle may have bumps at the left end, whereas for high tones the squiggle may have a prominence at the right-hand end. If the "lamps" on the display panel correspond to a typical reference spectra of phonemes *in an orderly fashion,* then each lamp could be assigned a *model* sound spectrum, and the actual recorded speech spectra could be compared against all the model sound spectra. The closest matches define the *responses* or the flickering on and off of particular lamps on the display. If the panel is also *labeled,* i.e. symbols indicate the corresponding phonemes, then one could via electronic means *type out* the lamp flashes in a sequence. A real speech recognizer would be much more complicated than this, because we must know how many samples or measurements from the continuous speech signal are collected in order to define exactly one phoneme. Furthermore, the transient phonemes need a separate analysis.

The process of creating a self-organizing map requires two layers of processing units: the first is an input layer containing processing units for each element in the input vector, the second is an output layer or grid of processing units that is fully connected with those at the input layer. The number of processing units at the output layer is determined by the user, based on the initial shape and size of the map that is desired. Unlike other neural networks there is no hidden layer or hidden processing units. Figure 0.1 shows a simple comparison between a supervised (e.g. backpropagation) neural network and a self-organizing (feature) map.

When an input pattern is presented to the network, the units in the output layer compete with each other for the right to be declared the *winner.* The winner will be the output unit whose incoming connection weights are the closest to the input pattern in terms of Euclidean distance. Thus the input is presented and each output unit competes to match the input pattern. The output that is closest to the input pattern is declared the winner. The connection weights of the winning unit are then adjusted, i.e. moved in the direction of the input pattern

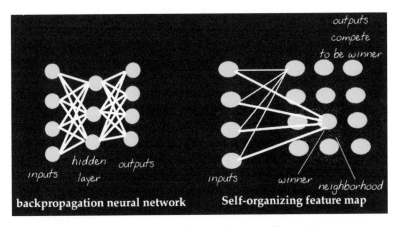

Figure 0.1 Backpropagation neural network versus self-organizing map.

by a factor determined by the learning rate. This is the basic nature of competitive neural networks.

SOM creates a topological mapping by adjusting not only the winner's weights, but also the weights of the adjacent output units in close proximity of the neighborhood of the winner. So not only does the winner get adjusted, but the whole neighborhood of output units gets moved closer to the input pattern. When starting from randomized weight values, the output units slowly align themselves such that when an input pattern is presented, a neighborhood of units responds to the input pattern. As training progresses, the size of the neighborhood around the winning unit decreases. Initially, large numbers of output units will be updated, but as the training proceeds smaller and smaller numbers are updated until at the end of the training only the winning unit is adjusted. Similarly, the learning rate will decrease as training progresses, and in some applications, the learning rate will decay with the distance from the winning output units.

The result is weights between the input vectors and the output neurons that represent a typical or prototype input pattern for the subset of inputs that falls into a particular cluster. The process of taking a set of high-dimensional data and reducing it to a set of clusters is called *segmentation*. The high-dimensional input space is reduced to a two-dimensional map. If the index of the winning output is used, it essentially partitions the input patterns into a set of categories or clusters.

A SOM also has the capability to *generalize*. This means that the network can recognize or characterize inputs it has never encountered before. A new input is assimilated with the map unit it is mapped to. Furthermore, even input vectors with missing data can be used to lookup or forecast the values of the missing data based on a trained map.

Simple Example: Mapping Scotch Whiskies

In this section we present an example based on a large data matrix of Scotch whiskies. The whiskies of Scotland fall into two categories: grain and malt. The malt whiskies are divided into four groups according to the geographical location of the distilleries. These are: the Highland malts, to be found north of an imaginary line from Dundee in the east to Greenock in the west; the Lowland malts, to be found south of that line; the Islay malts, to be found on the island of that name; and the Campbeltown malts, to be found in that town in the Mull of Kintyre. Each and every distillery in Scotland turns out a product which is unique and which has distinctive qualities that need to be recognized for any real appreciation of Scotch whisky.

In the *Malt Whisky Companion* (1989), Michael Jackson describes 109 different single malt whiskies produced in Scotland. For each of these whiskies he provides expert evaluations of the color (14 terms), the nose (12 terms), the body (8 terms), the palette (15 terms) and finish (19 terms). In addition, Jackson provides the age, the percentage of alcohol, the distance between distilleries (Scotland is divided into districts and regions), and an overall score for each whisky. These add up to 72 attributes on 109 Scotch whiskies.

We selected this data because it is

- a large and complex data set;
- described in "A Classification of Pure Malt Scotch Whiskies" (Legendre, 1994), which provides a benchmark for comparisons;
- a public domain data set (that can be downloaded from *http:/ alize.ere.umontreal.ca/~legendre*);
- an example that can keep the interest of financial analysts, economists, and marketing professionals alike.

We start by plotting the age, the percentage of alcohol, and the overall score of some 20 whiskies. A simple chart of three variables on 20 whiskies is shown in Figure 0.2. We used the left axis for the age and percentage of alcohol and the right axis for the overall score. We plot only a sample of 20 whiskies or 18% of the data because a chart with 327 (109 times 3) columns would not be very readable. This simple chart illustrates the problems we discussed regarding multi-dimensional data sets: it is simply not possible on the basis of a chart like this to select malt Scotch whiskies and take into account all the features or qualities that are available in this case.

Public domain software on the web, e.g. the R-package, allows us to create a *dendrogram* from this data, apply principal components analysis, *K*-means clustering, and even run an MDS. We will leave these

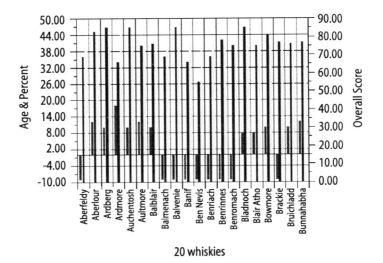

Figure 0.2 Age, percent alcohol and score of 109 pure malt whiskies.

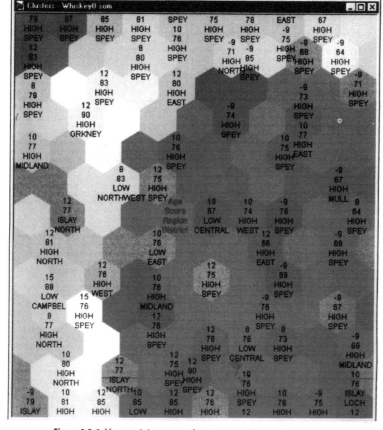

Figure 0.3 Self-organizing map of 109 pure malt Scotch whiskies.

as exercises for the reader. The limitations of each of these techniques has already been discussed. So, let's explore what a self-organizing map approach could do.

If the entire original data matrix is submitted to a SOM, we obtain a "display" like the one shown in Figure 0.3. In this figure each hexagon represents a node on a map or a processing unit of the output layer of this neural network. The shade of each node on this map indicates a pattern among all of the attributes used for this map.

To illustrate the differences we labeled the nodes on the SOM map with the age, overall score, and the location of the distilleries on the nodes of the map. The map clustered 109 whiskies into about nine clusters of which seven contain more than one node. The largest cluster can be found on the right side of the map. Most whiskies in this cluster are relatively young, contain about 40% alcohol and have overall scores between 70 and 74. Another cluster located in the middle of the map, contains whiskies that are on average about seven years old, contain 40% alcohol, and have an average overall score around 80. In the top left corner of the map we find whiskies that are older, have more alcohol content, and attain an overall score in the high 80s or low 90s. While the labels on this map show only a few attributes, the SOM has taken into account all 72 attributes in the data and has identified data groupings based on the entire data set.

Figure 0.4a–e shows three-dimensional representations of the differences in color, body, nose, palette, and finish among these single malt whiskies. Each of these figures focuses on one set of attributes. In each figure, the x-axis shows the clusters formed by the SOM; the y-axis shows the original terms used for assessing the quality of the malts (e.g. yellow, very pale, pale, gold for color); the z-axis shows the relative contribution of each attribute to each cluster.

The same patterns can also be extracted from a map of Scotch whiskies with the displays of the individual attributes used in the clustering. Figure 0.5, for example, shows three displays of individual attributes. The grayscale at the bottom of each panel goes from very small to very large values for each attribute. For example, the overall score for whiskies goes from 64 to close to 90. The lowest values can be found on the right side of the overall score panel and in consequence contribute to the main cluster found on the right side of the SOM map. The highest overall score can be found at the left top of the panel and thus create the cluster of best whiskies in the top left of the SOM map. The SOM map can thus best be interpreted by looking at the individual panels and reading the values from their scale.

This map, based on 72 attributes of 109 single malt whiskies, provides a very simple illustration of the power of self-organizing maps. For example, looking at the lowest values of the age, that is whiskies that are less than four years old, we note that they all belong to a main cluster in the right lower corner of the main map. From that map we can read which distilleries and in which districts and regions of Scotland those younger whiskies come from. Of course, if one is only interested in age, a simple sorted listing could have provided the locations of their origin. If one is interested in age, alcohol level, and price (which is not included in this data set), one could even consider an optimization which could produce the oldest whisky with, say, the highest alcohol level at the lowest price.

If one is, however, interested in finding the most typical single malt whiskies produced in Scotland,[1] then a simple sort or optimization would not allow to obtain the different types. With SOM one can produce a map that shows the most typical single malt whiskies based on dozens of attributes, use this map to select the most typical whiskies, and organize a whisky-tasting evening!

---

[1] On a visit to Dunedin for the ICONIP'97 meeting we learned that the Wilson distillery in Dunedin in New Zealand is the only one authorized outside Scotland to label its products 'Scotch whisky'.

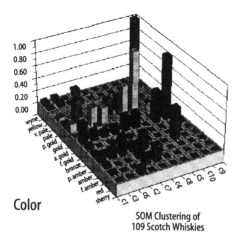

Color

SOM Clustering of
109 Scotch Whiskies

Figure 0.4a

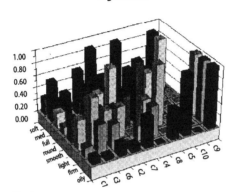

Body

SOM Clustering of
109 Scotch Whiskies

Figure 0.4b

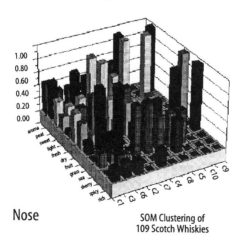

Nose

SOM Clustering of
109 Scotch Whiskies

Figure 0.4c

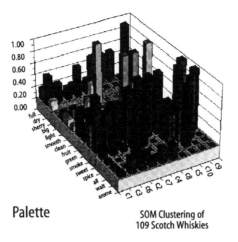

Palette   SOM Clustering of
                           109 Scotch Whiskies

Figure 0.4d

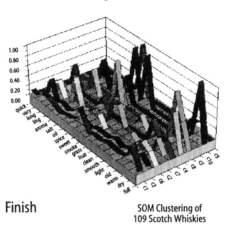

Finish   SOM Clustering of
                           109 Scotch Whiskies

Figure 0.4e

Overview

The core of this book is about data clustering, visual representations, and the discovery of new patterns and relationships in large multi-dimensional data sets. New patterns and relations can be used to make financial projections, to estimate market values, to rate financial instruments, to select investments in mutual funds, stocks or bonds, to undertake risk and portfolio management, to identify market potentials, to segment customers, to track which customers are buying what, when and the like.

This book contains more than two dozen applications of self-organizing maps. The main focus is on financial applications. There are also several real estate applications, one example of the use of SOM for the analysis of consumer preferences, and one example on the use of SOM for the organizing of large document collections. Other

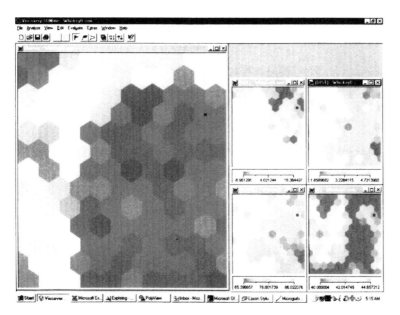

Figure 0.5 SOM and component planes of Scotch whiskies.

applications in engineering, psychology, image processing and other fields can be found in the references provided in the bibliography.

We have organized this book so as to facilitate learning by examples. Through many applications contributed by various authors we show how structures can be detected in large data sets, how multi-dimensional data can be mapped into two-dimensional displays. After these real-world applications we focus on the SOM methodology, the available software, and the best practices for creating maps.

The organization of this book is geared to readers who learn best from associations with real-world problems. Definitions and how the SOM method works are provided only after real-world applications are discussed. Readers who learn best from examples should first read the overview of applications that follows and then proceed by browsing through one or more applications. Readers who prefer to learn first the details of the methodology can turn immediately to Part 2 and browse the applications after it. Those interested in an in-depth study of the SOM methodology can find more details in the second edition of *Self-Organizing Maps* (Kohonen, 1997) as well as many other papers that have been written about this approach.

This book is divided into two parts. Part 1 is on financial applications. Chapter 1 provides an introduction to a wide variety of financial applications. Carlos Serrano-Cinca shows five applications of SOM using financial data. These include (i) an analysis of financial information for corporate strategy formulation; (ii) visual diagnosis of the financial situation of companies; (iii) establishment of bond ratings; (iv) an analysis of the economic convergence of European countries; and (v) financial maps for decision support.

In Chapter 2 Eric de Bodt has his co-workers show how maps can be used for projecting the long-term evolution of interest rates. Since a Monte-Carlo procedure does not produce the classical mean-reverting properties of interest rate structure dynamics, Eric de Bodt proposes to use SOM as an alternative approach. Based on a Monte-Carlo simulation he constructs the conditional probability distributions of interest rate structure shocks. This procedure is not only able to produce interest rate structure scenarios which are stable on a long-term horizon but also exhibits properties compatible with the historical interest rate structure evolution used to compute the conditional probability distributions. To understand the behavior of the proposed approach, Eric de Bodt uses a four-step methodology.

Chapter 3 shifts from interest rate predictions to portfolio management. Guido Deboeck shows how self-organizing maps can be used to translate multi-dimensional mutual fund data into simple two-dimensional maps that can be used for selection of mutual funds and managers. Using data published by Morningstar Inc. he explores patterns among mutual funds. Several variables from the Morningstar database are used as inputs. The emphasis is on how SOM simplify classification of funds; how SOM creates more meaningful decision-support tables than simple sorted lists based on single or multiple criteria. The approach presented demonstrates how information on mutual funds can be displayed, how benchmarks can be constructed, and how improved indices can be derived for comparison of the performance of mutual funds.

In Chapter 4 Kimmo Kiviluoto analyzes the failures of small and medium-sized enterprises. In this study, conducted in co-operation with Kera Ltd., a service company that specializes in financing small and medium-sized enterprises in Finland, the phenomenon of corporate bankruptcy is examined. The author shows several methods for classifying data from financial statements.

Serge Shumsky and A. Yarovoy write in Chapter 5 from Moscow about the structure and failures of the Russian banking system. Using a two-layer self-organizing neural network they demonstrate how newly published data on Russian banks can be reduced and visualized for data analysis. Through coloring of the SOM maps they produce a SOM atlas of Russian banks.

In Chapter 6 Guido Deboeck discusses maps for investing in emerging markets. Based on data from some 30 emerging markets, maps are created and deployed for facilitating investments in emerging markets. These maps demonstrate how it is possible to detect patterns in fundamental and technical data of markets in Latin America, Asia and Eastern Europe, as well as in the evolution of emerging stock markets over time. In addition, comparisons of individual companies in banking, telecommunications, and construction businesses from around the globe are mapped. Maps show similarities and dissimilarities between companies. Chapter 6 also explores the strategic importance of using maps for asset allocations.

In Chapter 7 Marina Resta integrates the best properties of self-organizing maps with genetic algorithms. She shows how a hybrid algorithm can be applied for financial forecasting and the development of trading strategies. The results evaluated on the basis of financial criteria demonstrate that this integrated hybrid approach performs better than traditional SOM alone. Chapter 7 provides experimental proof of the value of neural networks and is a nice compliment to the evidence presented in *Trading on the Edge*.

In Chapter 8 Eero Carlson uses self-organizing maps to analyze real estate investments, in particular the value of land properties. Using data from the National Land Survey of Finland, he includes in his analysis geographic location data. He combines SOM with a Geographic Information System and pays special attention to scaling of the components by which the resolution and topology of a map can emphasize particular aspects. Sensitivity analysis is used for changing a few components or for focusing on the properties to be appraised. Carlson also outlines how to use a moving time window for introducing time and how to fine-tune the organization of maps with new observations.

Anna Tulkki describes in Chapter 9 the application of self-organizing maps for the appraisal of buildings. The most common methods used to analyze the value of real estate properties are econometric models. However, these have some weaknesses that make it difficult to form good, reliable models. First there is the linearity assumption, and then the problem of correlating variables. These make it tough to make good models of the value of real estate properties. In particular, the changes in real estate values and prices in the 1990s have shown the weaknesses of econometrics models.

In Chapter 10 Bernd Schmitt and Guido Deboeck analyze the preferences and attitudes of Chinese consumers. Using data derived from a recent survey conducted in Beijing and Shanghai in the People's Republic of China, the authors show how consumers in these cities differ: how their preferences and attitudes differ in regard to brands, and foreign and new products, and how various groups of consumers can be identified. This chapter also contains some novel findings on how the living and dining habits of people in Beijing and Shanghai can be segmented into a small number of clusters.

Part 2 is about SOM methodology, software tools, and techniques. In Chapter 11 Teuvo Kohonen provides an in-depth overview of SOM and shows step by step how the self-organizing algorithm works. An important novelty is the addition of a sort section on batch SOM.

In Chapter 12 Timo Honkela et al. describe an extension of SOM to handling textual material in large document collections. In the rapidly growing worldwide web there is a vast amount of useful information available, but reaching it is not straightforward. Several search engines of the WWW provide keyword access to websites. Often these engines result in massive references with low accuracy in regard to the Boolean expression of the original request. Considerable efforts have been made to develop alternative methods. Honkela presents a method for full text-information retrieval, called WEBSOM. The WEBSOM method is based on the SOM algorithm. Several examples are mentioned and are also accessible on the web.

In Chapter 13 Guido Deboeck discusses software programs to implement self-organizing maps including the SOM_Pack prepared by the Neural Networks Research Center of the Helsinki University of Technology; the SOM Toolbox implemented in MatLab, and Viscovery SOMine, a commercially available tool for data mining with SOM that provides excellent visualization capabilities.

In Chapter 14 Samuel Kaski and Teuvo Kohonen share important tips for processing of self-organizing maps as well as for automatic color coding of maps. Here we find expert advice on the selection of the map size, the shape of the map, scaling of the input variables, selection of the neighborhood function and learning rate, and the best initialization of the code book or model vectors.

Finally, in Chapter 15 Guido Deboeck outlines a step-by-step procedure for the development of SOM applications. The implementation of this procedure is illustrated with an example on country credit risks. This chapter also summarizes the best practices or lessons learned from applying self-organizing maps in the many applications presented in this book.

While this review of financial applications covers a lot of ground, it is clear that after browsing through them you may conclude that a lot remains to be done! We encourage you to think about the various approaches presented here, to enhance these approaches and to adapt them to other domains. We encourage you to try out the SOM software packages discussed in this book and to create innovative applications of self-organizing maps. If you send us a copy of your work, we will consider it for inclusion in future revisions of this book.

Part 1

Applications

1 Let Financial Data Speak for Themselves

Carlos Serrano-Cinca

Abstract

Carlos Serrano-Cinca of the University of Zaragoza in Spain discusses five different applications of unsupervised neural networks and self-organizing maps (SOM) using financial data: (i) analysis of financial statements and information for the formulation of corporate strategy; (ii) visual diagnosis of the financial situation of companies; (iii) establishment of bond ratings; (iv) analysis of the economic convergence of European countries using macro-economic indicators; and (v) self-organizing maps as decision support systems. The wide variety of financial applications presented by Carlos Serrano-Cinca shows that SOM is an important tool for initial data analysis.

1.1 Initial Analysis of Financial Data

Financial analysts use a variety of techniques and tools to convert data into information that is used in decision-making processes. When performing this task, financial analysts make use of different techniques. The availability of different techniques has developed in parallel with the evolution of scientific disciplines such as statistics, operational research and computing.

For example, browsing through the menus of any statistical computer program shows the large number and variety of techniques available to financial analysts. Such programs bring advanced methods of numerical calculus and complex forecasting models within the reach of anyone. Despite the growing complexity of these statistical tools, simple statistical analysis and a few graphics can make a significant contribution to the process of converting data into useful information. At the very least, simple statistical analysis and a few graphics prepare the way for a more thorough analysis. A glance at the data can translate huge volumes of data into manageable information. In consequence, exploratory methods could become more and more important in the analysis of financial information.

According to Chatfield (1985), the initial examination of data or initial data analysis (IDA) is a valuable stage of most statistical investigations, not only for scrutinizing and summarizing data, but also for model formulation. It is often the case that IDA is sufficient, obviating the need for more sophisticated models. Chatfield feels these methods are generally undervalued, often neglected, and sometimes actively regarded with disfavor. Many academics reject these analyses, considering them to be "ad hoc" or trivial; professional analysts and managers appreciate them to the extent that they are included in any executive information system because of their intuitive character.

Of all the IDA techniques, the most popular ones are the univariate methods, which treat each variable independently. Examples include histograms, stem-and-leaf plots and box-and-whiskers plots. These graphics provide complete visual information on the data and are extremely valuable exploratory techniques. For large groups of variables, multivariate techniques must be employed. Multivariate techniques include cluster analysis and multi-dimensional scales which were introduced earlier.

Cluster analysis (CA) can be used in situations where we are interested in grouping together a body of patterns where a priori we have no clear idea of their relationships. CA can group the initial data into various groups and sub-groups, which are represented graphically in a figure known as a *dendrogram*.

Multi-dimensional scaling (MDS) attempts to produce geometrical representations of the data set. Starting from a distances table, it graphically represents the points in a series of maps that summarize the main features of the data and are easy to interpret.

Self-organizing map (SOM) is a complementary technique that can be used for initial data analysis. SOM applies a process known as self-organization to the initial data set. This processing allows the initial data analysis of financial and economic data to speak for themselves. For example, when financial data on a group of companies is introduced, these companies will be self-organized in such a way that those with similar financial characteristics will be located close to one another on a map. Additionally, SOM allows us to study the evolution of a company over time (by introducing information coming from different accounting periods) to place a company in relation to its competitors, and to prepare sectorial maps.

In this chapter we show how to apply SOM to diagnose company solvency, to establish bond ratings, to formulate or detect a company strategy in relation to the sector in which it operates based on published accounting information. We will also show how to compare financial and economic indicators of various countries. Naturally, allowing the data to speak for themselves does not mean that such analysis is sufficient: all initial data analysis is an important first step within any empirical research; it is no more than a first step and it must be completed with other analyses. In this chapter we shall demonstrate the integration of SOM into a decision-support system used for predicting the probability of company bankruptcy. Employing SOM does not imply that the use of other well-known techniques is renounced. Indeed, we will compare SOM with multivariate statistical models such as Linear Discriminant Analysis, as well as with neural models such as the Multilayer Perceptron.

1.2 SOM as a Tool for Initial Data Analysis

This section describes a number of practical cases of the use of the SOM as a tool for IDA. In each case, we will set specific objectives, namely the classical applications for the external analysis of quantitative financial information. After discussing the data, we shall apply SOM for IDA. In these examples we have employed all available variables and cases, without carrying a prior selection of the variables or discounting patterns that might present atypical values. The only change that has been applied to the financial variables is their standardization to mean zero and variance 1, in order to avoid problems that may occur when variables are measured on different scales.

The Euclidean distance is used for the similarity measure. Although it is possible to think of many ways of comparing individual companies, the easiest is to calculate the Euclidean distance between them using standardized ratios as variables. Any two companies with very similar ratio structures will show a small distance between their standardized ratios and thus will appear closer to one another on the SOM. The converse will also be true: if two companies have very different ratio structures, the distance between their standardized ratios will be large.

As we will see, in some cases the view provided by the maps is almost sufficient enough for the results to be interpreted. However, we must keep in mind that the SOM is not a panacea and, therefore, we should always complete this type of analysis with other tools. There are many other traditional statistical techniques that are more suited to solve some aspects of a problem, namely univariate or multivariate analysis, and the best method for solving each particular aspect should always be chosen.

1.2.1 From Financial Information to Corporate Strategy Formulation

This first case studies the strategic positioning of the Spanish Savings Banks on the basis of their published financial information. The Savings Banks play a very important role in the context of the European Union. Their market share is approximately 25% of the external funds of the financial system. This percentage is even higher in Spain (43%) and has not ceased to increase in the last few decades, with its total now reaching some 25,000 million US dollars. This sector is not only important, but also highly successful, when account is taken of its profitability, its immunity from crisis, and the solid image it offers to its customers. The sector is dominated by a small number of institutions. Attempts have been made to break down barriers to competition, particularly in the framework of the European Union (EU). For this reason it has been targeted by competitors from other countries, and many foreign institutions have set up offices in Spain. The Savings Banks are now immersed in a merger process in which many of them are involved. They will have to meet new challenges, and in the context of EU construction, many questions can be posed with respect to their future.

One of the most interesting approaches that can be adopted when investigating a sector is the concept of the strategic group. This can be defined as a group of firms in an industry, which have many similarities in their cost structure, levels of diversification and systems of organization, as well as the provision of incentives. The strategic group is a unit larger than the firm but smaller than the sector. Each sector can have different strategic groups within it, depending upon the strategy followed by the companies which make it up. A knowledge of the strategic groups within a specific industry is useful for the companies. If a particular company wishes to change its strategic positioning, it must have a prior knowledge of the fundamental problems it will have to face when designing an appropriate plan of action. This knowledge might also be useful for those companies who are considering entering this sector; to evaluate its interest or attraction, to know with greater certainty the opportunities for future profits, and to take advantage to the greatest extent possible of structural changes that might take place in the sector.

In our example, the strategic groups have been obtained using data drawn from the financial information supplied by the Spanish Savings Banks. The data employed in the study is taken from the Statistics Yearbook of The Spanish Confederation of Savings Banks in its Annual Report on the results of the sector, and it corresponds with the public information on each entity for 1991. We have used 30 financial ratios, which attempt to capture profitability, capital structure, financial costs, risk structure, etc. The companies analyzed are the 56 Savings Banks that were operating in Spain in 1991. A SOM model has been applied in the search for the self-organization of these Savings Banks. Figure 1.1 shows how Saving Banks are situated on a map after the training.

An examination of this map provides powerful insights into the strategic groups of the sector. Note that the Savings Banks are distributed throughout the map according to the geographical area in which they operate. The three Basque Savings Banks are located in the upper part, mirroring their real location in Spain. Various Catalan, Valencian and Balearic Savings Banks are grouped to the right. The central zone corresponds to the area of Castile, with Madrid in its center, whilst the Andalucian Savings Banks are found to the south. Although not occupying the geographical position which corresponds to them, the two Savings Banks from

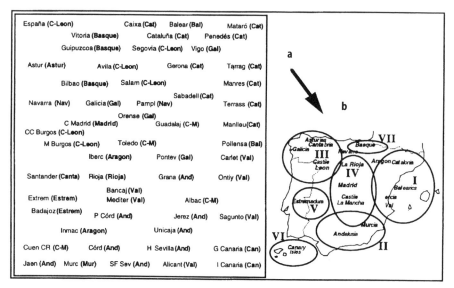

Figure 1.1. a The Spanish Savings Banks have self-organized according to their geographical distribution. **b** Strategic groups.

Estremadura and, in turn, those from the Canary Isles, appear close to each other on the SOM. Bearing in mind that the data does not contain any information of a geographical type, it is truly surprising that its self-organization results in an image that is very similar to the map of Spain.

Cluster analysis has proved to be useful in delimiting the groups. When studying the results provided by CA, the SOM results were confirmed and detailed with greater precision. Seven strategic groups have been obtained, with these corresponding to a further seven Spanish macro-regions, to which the following names have been given:

1. Crown of Aragon;
2. Al Andalus;
3. North-West;
4. Center;
5. Estremadura;
6. Canary Isles;
7. Basque.

Naturally, the correspondence between geographical position and strategic group is not exact for all the Savings Banks.

Why have the Savings Banks been grouped together in this manner? Spain is divided into 50 provinces, grouped into 17 Autonomous Regions, with each region enjoying a certain degree of self-government. This regional division, although somewhat recent in that it flows from the Constitution of 1976, tries to reflect the traditional division of Spain into natural and historical regions. The country divides itself into a number of natural regions; at least four different languages are spoken in Spain; there are significant differences between the North and the South, as well as differences in salary and in cultures. Furthermore, all the Autonomous Regions have their own jurisdiction over matters pertaining to their Savings Banks. However, none of these factors have been introduced into the SOM, which has only been supplied with data taken from the Balance Sheets and Profits and Loss Accounts of the Savings Banks being analyzed. What this means is that the Savings Banks belonging to these strategic groups which, as can be seen, coincide with the Regions, follow the same strategy and, as a result, have similar margins, profitability, levels of bad debt, solvency and productivity ratios, a similar financial structure, etc.

The strategic groups detected by way of this empirical study have revealed the importance of the regional component. According to many specialists and Savings Banks Managers, it is this strategy of "territorialism", maintained up to now by almost all of them, that has been the differentiating factor which has contributed to the growth and strengthening of this group of financial entities. Territoriality is the key factor from which the Spanish Savings Banks draw their strength. Their social objectives, as non-profit-making entities, another of their distinctive factors, is also closely linked to this factor.

In order to complete the study, we need to know the strategy followed by the Savings Banks belonging to the different strategic groups. To that end, we have calculated the average value of each one of the financial ratios for each strategic group and then carried out a bivariate analysis of ratios which has assisted in identifying the features corresponding to each group. The values of each pair of ratios have then been represented in graphic form.

Figure 1.2 shows an example of two of these graphics, illustrating Financial Margin versus Economic Profitability and Employee Productivity versus Personnel Expenses, respectively. It further shows the level of technology employed, as illustrated by the number of auto-tellers over the number of employees.

For the sake of brevity we do not propose to describe the bivariate study of ratios, although we shall give some consideration to the results. We have given the name "Leader" to the strategy followed by the Savings Banks of the Center zone, because these are the most productive, solvent and technologically advanced. "Dolce vita" can best describe the strategy followed by the Estremadura Savings Banks, as they enjoy very good results, and maintain very high margins. For the Basque group, the strategy has been called "Europe", on the basis of their low margin, high productivity, and the highest personnel costs, which we put down to a high degree of competition. "Moderation" is the key word to identify the strategy followed by the North-West group. The Crown of Aragon group occupies an intermediate position between the Basque and Center groups, and we have called its strategy "Resistance". For the Al Andalus strategic group, with high financial margins, low profitability and a lot of bad debts, the name we have chosen is "Risk". Finally, for the Canary Isles group, their strategy can best be described as "Isolation", in that this group presents singularities which derive from its island nature, with scarce competition from other banking institutions.

The study was completed with information from the year following the chosen year of study and here, as a special circumstance, the database was incomplete because two of the ratios could not be supplied. Incomplete data is a serious inconvenience in many analysis techniques. It is not so for the SOM, which simply calculates the distances between the firms without taking the value of the lost ratio into account. In this study the displacement of certain Savings Banks, which are

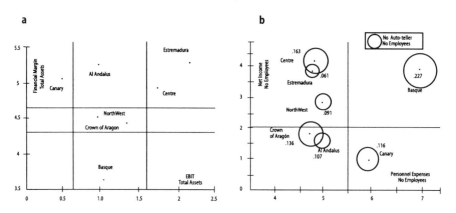

Figure 1.2. A simple example of the bivariate analysis of ratios. **a** Financial margin versus economic profitability. **b** Employee productivity versus personnel expenses.

modifying their strategy, could be observed. The policy of expansion of Spanish Savings Banks into other Autonomous Regions, which has not always met with success, is currently the subject of debate. It is probably the regional differences that are so characteristic of Spain that have not properly been taken into account. In this context, the foreign financial entities, most of them from the EU, that are opening branches throughout the country should themselves be aware of these peculiarities of the Spanish financial sector, which are nothing more than an extension of the rich cultural, linguistic, geographical and social variety of present-day Spain.

1.2.2 Visual Diagnosis of the Financial Situation of Companies

In this section we describe the capabilities of SOM to produce an image of the financial state of a wide variety of companies. The basic objective is to analyze their solvency. The SOM will be used to make a visual diagnosis of the situation of a variety of companies belonging to the same sector. In the earlier example of the Savings Banks, these entities were grouped together according to the similarities in their economic-financial structure. The same is the case in this example, but with the difference that we start from a sample which contains companies in financial difficulties and others that are solvent. By way of the self-organization of this sample, our aim is to group companies in difficulties together in such a way that it is possible to distinguish them from the healthy companies.

The data refers to the Spanish banking crisis of the early 1980s, following the work of Martin del Brio and Serrano-Cinca (1995) and has been taken from the "Private Banks Statistical Yearbook" published in 1982. The sample contains information on 66 banks, 29 of them bankrupt. From amongst the many financial ratios available we have used nine, selected by means of linear regression because of their statistical significance. Figure 1.3 shows the results of the self-organization of these banks.

The unsupervised network discovered similarities between patterns and has clustered similar patterns together. We can see how the banks that went bankrupt have located themselves to the right of the map, whilst the solvent ones are located to the left. Note that we have not provided information about the solvent or bankrupt state of the banks. The network itself, without any supervision, has found features in the input data that differentiate both situations. We call these Self-Organizing Solvency Maps, as discussed earlier in Martin del Brio and Serrano-Cinca (1993). An interesting feature that can be seen in Figure 1.3 is the clustering of The Big Seven, the seven biggest Spanish banks (banks 60 to 66 in the figure). Following the terminology of Section 1.2.1 these seven banks form a true strategic group, characterized as large-scale, very influential, solvent banks, although not the most profitable.

The advantage of a graphical representation is that a clear explanation can be given for the classification of a bank into the bankrupt or solvent set. In the case of the banks that form part of the analysis, a clear picture does emerge. By means of SOM we can see the financial situation of a bank in a particular year: by showing its pattern to the trained network and looking at the more activated neurons on

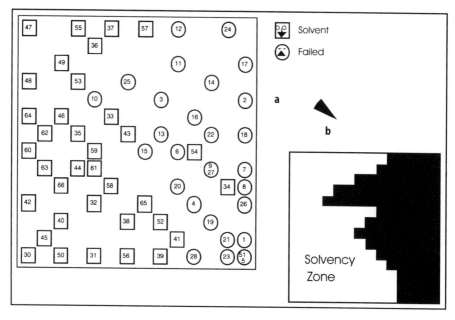

Figure 1.3. a Numbers 1 to 29 are banks in financial difficulties that subsequently went bankrupt. The self-organization of these banks has discriminated between the solvent and the bankrupt. **b** From **a** it is possible to distinguish two areas in the so-called self-organizing solvency map.

the map. If these are located in the crisis zone, the bank is in a critical situation; if they are in the solvent area, then it is solvent.

One of the first applications of SOM was developed by Kohonen (1989, 1990) himself, namely the phonetic map, which was obtained by training a self-organized map with Fourrier transforms of the phonemes of a language. When a word is pronounced, the neurons are stimulated in a sequence that is represented in a trajectory. Figure 1.4a shows the trajectory obtained when pronouncing the Finnish word "humppila". Connecting this system to a microphone allows us to use SOM as an automatic voice recognition system.

We can do something similar with our solvency map, namely observe the evolution of a bank, by introducing the financial information from various years. If a bank clusters with the failed banks, then it must be treated with care, on the grounds that its financial structure is not different from that of other banks that have failed in the past. If it clusters with non-failed banks, that concern disappears. In Figure 1.4b we show the evolution of a particular bank, the "Banco de Descuento" over a period of 18 years. Note that between 1973 and 1976, the bank is located in the solvency zone. In 1977 it is introduced into the bankruptcy zone and three years later, in 1980, it went bankrupt. After several years, during which it was technically bankrupt, it was reorganized and in the map it is introduced once again into the solvency zone. However, in the final years of the 1980s it returns to the bankruptcy zone and it did indeed go bankrupt a second time in 1990.

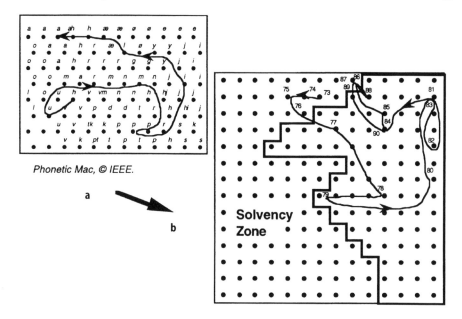

Figure 1.4. From the phonetic map to the solvency map. Time evolution of a bank up to its bankruptcy.

When a bank is located in the frontier region between the bankruptcy and solvency zones we must pay particular attention. The advantage of the SOM representation is that this region is quite evident. However, given their importance these zones must be specified with the greatest possible precision. In the next section we will discuss bond ratings, but will return to the integration of SOM into a decision-support system and the use of various multivariate mathematical models to study business solvency in the last section of this chapter.

1.2.3 Evaluating Financial Risks and Establishing Bond Ratings using SOM

Another of the classical applications of financial analysis is the evaluation of the financial risk of bond issues. The objective of this section is to develop a SOM that is useful for potential investors. We will concentrate on the Spanish bond market. According to Standard and Poor (1995) the Spanish bond market is the fourth in size in the EU in terms of private debt, after the United Kingdom, France and Germany. The evaluation of credit risk is rapidly becoming an important issue, particularly after recent cases of bankruptcy, some of them involving well-known issuing institutions.

Bond ratings represent the opinion that specialized firms have of the relative ability of an issuer or lender to meet its interests or repayments promptly. Such ratings are, therefore, a useful service which could be of relevance when making investment decisions, since it provides information on risk. Even if the ratings are mainly a service to the investor, their existence benefits all the agents in international financial markets: issuers, financial intermediaries and market regulators. Financial risk is measured by means of a symbolic scale reflecting the level of risk, both long-term and short-term, which is associated with an issue of debt. The higher the risk of default on the bond, the lower the rating given.

Bond-rating companies claim that financial and non-financial information are both taken into consideration when producing a rating. They mention factors such as the competitive position of the issuer, its business plan, its management strategy, its future investment plans, and any external circumstance that may affect its future solvency. They also mention that financial statement analysis plays an important part, although they accept that no clearly defined criteria exist on how this is taken into account. Attempts have been made to predict bond rating from financial and non-financial information using statistical models; some examples are Pogue and Soldofsky (1969), Pinches and Mingo (1973), and Ang and Patel (1975). These studies have attempted to test the predicting ability of a small number of explanatory variables. For this reason, the most frequently used tools have been multiple regression and discriminant analysis. More recently, artificial neural networks have been employed for this purpose, see for example, Dutta et al. (1994), Singleton and Surkan (1995), and Moody and Utans (1995).

We will study the Spanish banking sector in 1993. In that year, it contained 167 institutions. Banks with only one or two offices were removed from the sample, which led to the exclusion of all non-EU foreign banks, although Spanish subsidiaries of foreign banks were kept in the sample. This process reduced the sample to 88 institutions, which accounted for 99.03% of all deposits. We have tried to include variables that reflect the whole spectrum of financial aspects which would be of interest to the analyst. These have resulted in 24 financial ratios obtained from the Balance Sheet and the Profit and Loss Accounts, which attempt to capture profitability, capital structure, financial cost, risk structure, etc. The financial data used was published in the Yearbook of the Higher Banking Council of 1993, see Mar Molinero et al. (1996). The dissimilarity between any two banks was calculated as the Euclidean distance between standardized ratios.

Figure 1.5a shows the results of the self-organization. On the basis of this map, and with the support of our prior knowledge of the sector, we can make a number of observations. First, the zones are not as clear as in Section 1.2.2 where bankrupt and solvent entities were available to us. Thus, it will be necessary to complete the information provided by the SOMs with other external information that has not been used earlier. In Figure 1.5b we have superimposed the debt classifications given in 1993 by Standard & Poor's to various entities in the Spanish banking system. S&P ratings are given in the form of an ordinal scale ranking from A-1 for the best to D for the worst. Unfortunately, in 1993 S&P classified just a few institutions in Spain. Only A-1 and A-2 ratings were available. Within the A-1 category, a plus sign (+) has been added to the issues considered to be of a very high quality. Figure 1.5b shows how the banks classified with an A-1 are found on the upper right-hand side of the map. We can conclude that this is a zone of low financial risk.

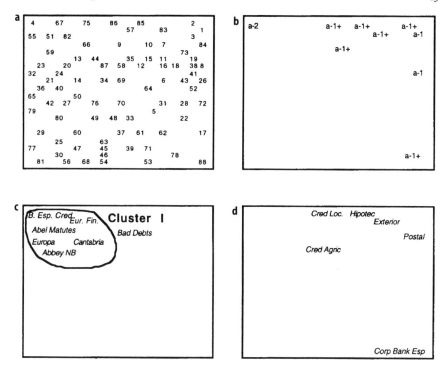

Figure 1.5. a Self-organized map. **b** Superimposed ratings given by Standard & Poor's. **c** Superimposed cluster analysis. **d** State-owned banks.

The only A-2 classification is the Banco Español de Crédito, located on the upper left-hand side. This bank was the subject of intervention on the part of the Bank of Spain on 28 December 1993 and this fact was already mentioned in the 1993 Annual Accounts. S&P rated it as A-1 under a negative credit watch in June 1993. This was lowered to A-2 just a few weeks before it was the subject of intervention. Moody's also lowered its rating. A detailed analysis of the banks in the upper zone of the map shows that their Balance Sheets reflect numerous bad debts. In order to delimit this zone with greater precision, we have complemented the SOM with CA. The results of the CA reveal the existence of a cluster made up of banks with serious financial problems located on the upper left-hand side of the SOM. In Figure 1.5c the banks which form this cluster, 4, 51, 55, 67 and 82, have been circled. Note how the Banco Español de Crédito (4) forms part of this cluster.

The use of SOMs as a support tool in bond rating can be employed when estimating the level of financial risk of those banks that have not been classified by agencies such as S&P or Moody's. Based on the recent history of the Spanish banks we can confirm that the model itself has been quite efficient. Cantabria (82), located towards the upper left-hand side of the map, and belonging to cluster I, has suffered a substantial amount of distress in recent times and was restructured under a new name in 1995. The Banco Abel Matutes (51), another bank in the same cluster, was restructured in August 1994 and merged with Catalá de Credit (32).

The only one of the 88 banks which has been removed from the Official Bank Register is the Banco de Inversión y Servicios Financieros (77). This took place in 1994. It can be found on the left of the map.

In Figure 1.5d we have superimposed a new piece of data, namely the six banks that were fully State-owned at that time. Four of these, Corporación Bancaria (88), Exterior (83), Hipotecario (85) and Crédito Local (86), have a rating, with all of them being given the highest ranking of A-1+. Note the position of Crédito Local, close to the cluster of banks with financial difficulties, as well as the atypical position of Corporación Bancaria. Given their position in the map, lower ratings should be expected. It is clear that the guarantee of the State is important for the rating agencies. However, these banks are currently in the middle of a privatization process, and it is expected that the State will limit its ownership to 25% of the shares. In fact, all four banks appeared under a negative perspective in July 1995, and the S&P agency explicitly stated that the cause of this outlook was the impending privatization.

1.2.4 The Economic Convergence of the EU Member States

In this section we will shift from the company and sector level to the macro-economic and country level. The construction of the EU is an historical event of the first order. One of the greatest concerns is to determine the economic conditions that are prevailing in the countries that will participate in this trans-national adventure. In this section of the practical application of SOM we analyze the similarities and differences between the countries involved in the construction of the EU. In order to acquire this knowledge we will compare the main economic-financial indicators of each country.

As a prior step we have selected a set of indicators that synthesize the economies of the countries involved. We have chosen as variables the macro-economic indicators that the European politicians agreed to include in the Maastricht Treaty. The spirit of this Treaty is that the economies of the EU member states should converge in such a way that integration is achieved without friction. To measure this convergence, the Treaty has fixed the following criteria. First, the achievement of a high degree of price stability, which will be apparent from rates of inflation. Second, the sustainability of the governments' financial position. This will be apparent from having achieved a government budgetary position without a deficit that is excessive. Third, the observance of the normal fluctuation margins provided for by the Exchange Rate Mechanism of the European Monetary System, for at least two years, without devaluing against the currency of any other Member State. Finally, the durability of convergence achieved by the Member State and of its participation in the Exchange Rate Mechanism of the European Monetary System, being reflected in the long-term interest rate levels.

As a source of information we have used the data corresponding to 1995 as provided by the European Commission. The dissimilarity between any two countries has been calculated as the Euclidean distance between standardized macro-economic variables. Figure 1.6 demonstrates the results of the self-organization of the countries using these variables.

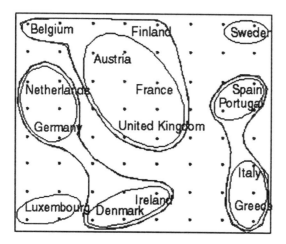

Figure 1.6. Self-organization of the 15 EU Member States, using the macro-economic variables proposed by the Maastricht Treaty. The results of a CA have been superimposed.

In order to delimit the regions with greater clarity and to interpret the results, we studied the synaptic weights. We obtained three maps, one with the maximum synaptic weights, one with the minimum and one with the greatest synaptic weights in absolute value for each neuron. Figure 1.7a contains the map which shows the greatest synaptic weights in absolute value. For each neuron of the map it indicates the feature, be it positive or negative, that has impressed it most. It is useful to determine both the aspects that each country must improve, as well as their strong points.

From both Figure 1.6 and Figure 1.7, we can note that Spain and Portugal, two countries that are geographical neighbors, appear close to one another in the SOM and also belong to the same cluster. The most obvious feature according to the synaptic weights map is that of inflation. Indeed, the reduction of inflation, then running at more than 4%, was fixed as the key objective for 1995 and subsequent years by both national Governments. Lying very close to these countries, both in the SOM and in reality, as fellow Mediterranean States, we find Italy and Greece. The most obvious feature for these two countries is the high rates of interest. It is perfectly understandable that the countries are self-organized according to their geographical location. There is clearly a Mediterranean Europe, made up of Spain, Italy, Greece and Portugal. In recent years this area has been characterized in economic terms by its higher levels of both inflation and interest rates. In the upper right-hand corner of the map we find Sweden, regarded as a model country in terms of social services, but which suffers from public deficit problems. Belgium, located in the upper left-hand corner, has its high level of public debt as a weak point. By contrast, and so far as strong points are concerned, Finland stands out by virtue of its low level of inflation, Austria for its low rates of interest, and countries such as Germany, Holland, Ireland and Luxembourg because of their low public deficit, with the latter even running a surplus.

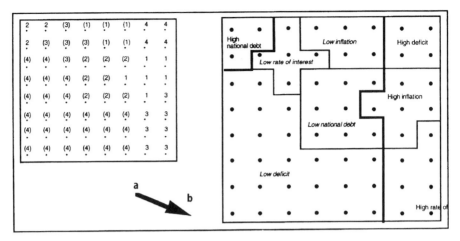

Figure 1.7. a Synaptic weights maps. It shows, for each neuron, which variable provokes the greatest response. The variables which stand out in the negative sense are shown in brackets. **b** We can determine which variable dominates over one or other zone of the map.

Ever since the signing of the Maastricht Treaty, the EU Member States have been making an effort to ensure that these indicators are met. Sometimes, the effort devoted to these macro-economic variables has been to the cost of other economic indicators, and the question is increasingly being asked whether what is being indicated by these variables is really so important. In our view, the situation of the real economy, of the companies that operate within it, and indeed of European citizens in general, is just as important as the convergence proposed by politicians. This then raises a number of further questions. Are the companies operating in the different EU countries sufficiently prepared so as to compete on equal terms? Would it not have been more important for governments to have been concerned with increasing the productivity and profitability of these companies, or increasing investment in research and technology?

In the second part of this case we use data on productivity, profitability, etc. in the EU Member States, obtained from the Balance Sheets and Profit and Loss Accounts of the companies of each country, before turning to the self-organization of these countries on the basis of the said data.

We find ourselves dealing with the analysis of financial statements in an international context and, in these circumstances, a prior requirement is the availability of homogenous financial information on the companies operating in each country. The differences in the accounting practices followed from country to country means that it is risky to compare the financial information of their respective companies. In our practical case, this information has been obtained from the BACH database, a project launched by the European Commission, which seeks to homogenize the aggregate financial data information of each country and place it at the disposition of users. This project is based in the Central Balance Sheets Offices, an official service which operates in each of the EU Member States. These offices are responsible for gathering the Balance Sheets and Profit and Loss Accounts of companies operating in various sectors, in order to analyze the economic-financial situation of these sectors in each country.

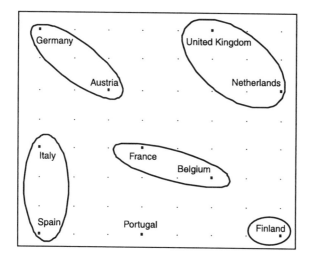

Figure 1.8. Self-organization of 10 EU Member States, using the financial information provided by the BACH database, with a CA superimposed.

The EU Member States participating in the BACH project in 1994 were Germany, Austria, Belgium, Spain, Finland, France, Italy, Holland, Portugal and the United Kingdom. We use the 16 financial ratios which form the BACH database. These ratios explain the economic results of industrial companies for each country in relation to the resources employed (gross profit, net profit and financial return), their relative costs (intermediate consumption, personnel costs and financial charges) and their financial structure (equity, indebtedness, debt structure and provisions). Figure 1.8 contains the SOM and, as in the earlier case, the results of a CA have been superimposed. Figure 1.9 reflects the synaptic weights map.

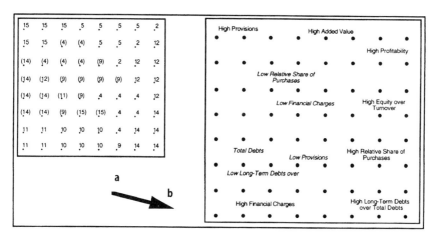

Figure 1.9. a Synaptic weights maps. **b** We can determine which variable dominates over one or other zone of the map.

As we can see, Spain, Italy and Portugal are located close to one another on the map and appear to form part of the same conglomerate. Their high financial charges stand out, which is coherent with the high interest rates shown on the map obtained using the Maastricht macro-economic variables. Furthermore, the strong proportion of short-term debt in their indebtedness ratios makes them very sensitive to changes in short-term interest rates. The opposite is the case with Belgium and France, which form a separate conglomerate. Finland is also characterized by the debt structure of its companies, where practically half of the debt is long term. Another conglomerate is formed by Holland and the United Kingdom, countries where high levels of added value and productivity can be noted. Ratio 15 includes provisions for pensions as its most important item. In this respect, companies operating in EU countries such as Germany and Austria accumulate important resources in their provisions. More than 30% of the total Balance Sheet of German companies is occupied with these items, whilst in Austria the proportion is 20%. The remaining countries rarely allocate funds to provisions for liabilities.

The latter map, which uses data taken from companies, is more important than the former, where the data is macro-economic in nature. What is more, *there can in the view of this author be no real convergence until we have a Europe that is a true reflection of its citizens.* This calls for the preparation of a new SOM and employing data which show the differences and similarities that exist between European citizens: including such indicators as the average wage, the consumption of wine, beer, olive oil and butter, the rate of unemployment, how leisure time is used, etc. *If we Europeans learn more about each other, we can converge in the positive aspects whilst maintaining our rich social and cultural diversity.*

1.3 Integrating SOM into a Decision-Support System

In this final section we turn to integrating SOM as a decision-support system. Predicting company failure is clearly one of the most important functions of financial analysis. Usually, the analysis of the financial state of companies is made from ratios obtained from financial data published by companies. Conventional methods based on univariate or multivariate analysis are applied to this data. Multivariate statistical models, such as discriminant analysis, logit, or supervised neural networks (such as the multilayer perceptron) are particularly worthy of mention because of their popularity. All these methods have in common that they seek to obtain a ranking that is easy to interpret as an indicator of company solvency.

In our view, whilst these rankings and indicators are useful, they are not free from drawbacks given the information they are based on. Various companies could have very different financial structures, and have nevertheless the same value for the solvency indicator. It is not easy to determine the financial features which characterize a company or the problems it faces on the basis of its solvency indicator alone, and it is thus necessary to complement it with other analyses.

We propose the use of the SOM to demonstrate the financial situation of companies in a graphic and intuitive form. As we have already demonstrated this in earlier sections, we go on to complete this analysis with other well-known techniques. Therefore, in this section the main objective is to develop a decision support system (DSS) for the prediction of corporate bankruptcy which integrates the SOM with other conventional methods based on multivariate analysis.

The database used in this study contains five financial ratios taken from Moody's Industrial Manual from 1975 through to 1985 for a total of 129 US firms, of which 65 are bankrupt and the rest solvent. This database was also employed in the work of Rahimian et al. (1993). In that work the sample was randomly divided into two groups, the first made up of 74 firms, used for training, and the second of 55, used for testing the models. We proceeded in the same way in our study, see Serrano-Cinca (1996). The ratios employed coincide with those selected by Altman (1968) in his pioneering work on the prediction of failure: r_1 = Working Capital/Total Assets; r_2 = Retained Earnings/Total Assets; r_3 = Earnings Before Interest and Tax/Total Assets; r_4 = Market Value of Equity/Total Debt; r_5 = Sales/Total Assets.

As in the earlier case where we studied bankruptcy in the Spanish banking sector, we first obtained the self-organized solvency map. Figure 1.10 shows the patterns which most stimulated each neuron. Note how this map has been obtained in a manner different from that of Figure 1.3, which was limited to showing the location of each company on the map.

It is possible to delineate two regions on the map with sufficient clarity, one corresponding to the solvent firms and the other to the bankrupt ones. However, the vision provided by the solvency map is not sufficient, in that we do not know

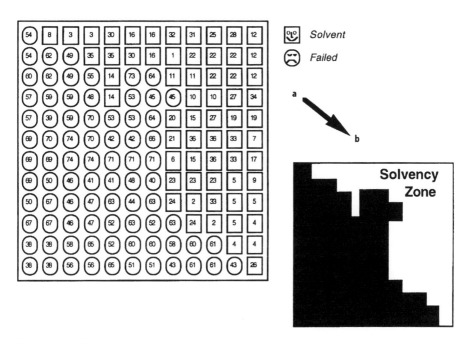

Figure 1.10. a The SOM. It shows, for each neuron, the firm which gives the strongest response. 1 to 36 are solvent firms; 37 to 74 contain information for one year prior to the incidence of bankruptcy. We can see two main areas, one consisting of neurons that have tuned to the bankrupt companies, and the other of neurons that have tuned to the solvent ones. **b** The solvency map.

how the grouping has been carried out, which variables have been the most relevant in the decision-taking process, etc. A study of the synaptic weights helps us to determine which variables dominate over one or other zone of the map. The map represented in Figure 1.11 indicates, with respect to each neuron, which variable has become specialized in recognition. That is to say, which positive or negative feature has impressed it most. From a study of the synaptic weights we can surmise a series of regions on the map: high earnings, low liquidity, etc. The upper right-hand zone of the map corresponds to high earnings ratios. The lower right-hand zone of the map are firms with high r_4, whilst the lower left-hand zone corresponds to low values of the four ratios. Ratios 2 and 3 contribute with greatest clarity to the delimitation of the bankrupt region. As expected, almost all the firms found in the bankrupt zone show patterns that are characterized by low earnings, whilst these are high for solvent ones.

We know that in the neural network the firms which are close to one another are firms which present similar patterns and that, by way of a study of the synaptic weights, we can obtain regions on the map. However, this might not be sufficient in order to clearly determine the frontiers between the firms. We also know that it is useful to complement the self-organizing solvency map with CA, and we have circled those groups over the map (see Figure 1.12).

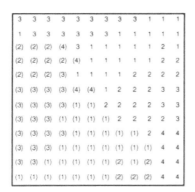

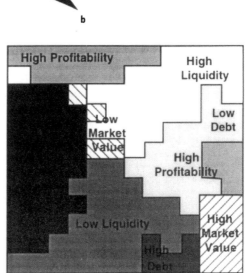

Figure 1.11. a Weight maps showing, for each neuron, which financial ratio provokes the greatest response, in absolute values. **b** Financial features.

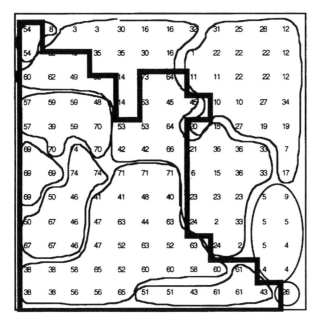

Figure 1.12. Superimposition of the three clusters onto the solvency map. The trace of the strongest line divides the plane into solvent and bankrupt firms.

From a study of the CA we can detect ten groups. Note how in the bankruptcy zone there are five complete groups, whilst in the solvency zone there are three. One group in the upper left-hand corner includes two companies, one solvent and the other bankrupt, inviting us to be prudent with the companies that are located here. Furthermore, there is a central zone that gathers both solvent and bankrupt companies, which appears to confirm the existence of an undetermined intermediate zone.

External information, which has not been used to create the SOM, can now be added in a structured way in order to explore its relevance. We can combine SOM with Linear Discriminant Analysis (LDA), the most popular mathematical model applied to the prediction of corporate failure. The objective of LDA is to obtain an indicator (Z score) which discriminates between two or more groups. An LDA has been carried out on the data used in this work. Thereafter we have obtained the Z score for each firm and then superimposed this indicator onto the SOM. This has allowed us to obtain some regions made up of firms whose solvency is similar according to the LDA. These regions have been given the name insolvency regions; these are four in number and can be seen in Figure 1.13a. Note that the two insolvency regions higher than 7 belong to the solvency zone and, similarly, how the insolvency region lower than 2 is included in the bankruptcy zone. Finally, the central zone groups firms with Z scores between 2 and 5.

Another neural model, the Multilayer Perceptron (MLP), can also be used to obtain the insolvent regions. This model has the common objective with LDA of obtaining a Z indicator which can be used as a measure of the solvency of the

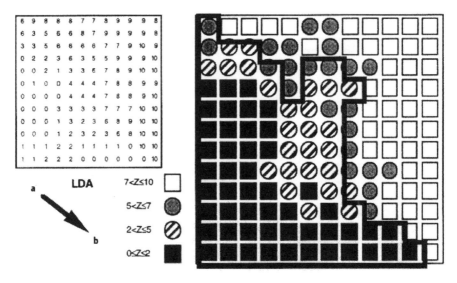

Figure 1.13. a Superimposition of the results of Linear Discriminant Analysis (LDA) on the solvency map. **b** Insolvent regions.

companies and is also capable of separating non-linear patterns. In fact, LDA is a particular case of the single layer perceptron.

With this DSS, and despite the complexity resulting from the combination of different tools, it is very easy for a user to evaluate the solvency of an entity by introducing no more than the values of its ratios. The model shows us, first,

Figure 1.14. Test firm number 1. The intensity with which the neurons are stimulated is indicated in different shades of gray. The figure shows a typical screen of the computer software developed for this exercise.

whether the firm is in the solvency or bankruptcy zone, and then, by studying the map of the regions, which financial features stand out. Furthermore, it shows us to which cluster the firm belongs and which firms present similar ratios. Finally, it shows us whether the firm belongs to one or other of the insolvency regions, according to LDA or MLP.

The usefulness of this approach is demonstrated by analyzing the first firm in the test data set. The ratios for this firm had not been used to train the network, thus no neuron in the output layer is expected to exactly represent it. However, those output neurons that are associated with firms which are very similar to the test firm will be strongly stimulated, whilst those that are associated with dissimilar firms will have low stimulation. We have introduced the ratios for this firm onto the SOM and we have obtained the map which appears as Figure 1.14.

The information supplied by the DSS is quite full and the neurons which are stimulated by its presence are various. The intensity with which they are stimulated is indicated in different shades of gray. Note that nearly all of them are neurons that are specialized in recognizing solvent firms. To be exact, the winner neuron is in the solvency zone which, from a study of the synaptic weights, is the zone with a very high ratio 2 and quite high ratios 1 and 3. With respect to the cluster analysis, the winner neuron is found in group III which includes the solvent firms. Furthermore, the three or four neurons which are in greatest synergy with pattern number 1 are in the zone of insolvency close to 10. On the basis of all this, we can conclude that it is a solvent firm.

As its principal advantage, this DSS provides a complete analysis which goes beyond that of the traditional models based on the construction of a solvency indicator, also known as the Z score, without renouncing simplicity for the final decision maker.

2 Projection of Long-term Interest Rates with Maps

E. de Bodt, Ph. Grégoire and M. Cottrell

Abstract

There are many models for pricing financial assets. Their goals are usually to reproduce the price of quoted financial assets, e.g. the prices of bonds or to make projections and thus provide evaluations of the risk exposure of assets portfolios. A risk-management policy can be chosen and hedging ratios decided using models. More than the theoretical evaluation of financial assets, determining the hedging ratios is often the essential goal of those models. Two main approaches exist for modeling the interest rates structure and its dynamics: a parametric approach and a non-parametric one. The non-parametric approach developed in this chapter does not assume a priori hypotheses on the functional form of the process generating the interest rates structure and on the form of the distribution that characterizes the dynamic of the random variables observed. Using a historical data set, Eric de Bodt, Philippe Grégoire and Marie Cottrell use the SOM algorithm to approximate both the distribution of the interest rates structure and its deformations over time (the shocks on the interest rates structures). On this basis, a Monte-Carlo simulation gives long-term interest rates structure evolution on a 5-year horizon.

2.1 Introduction

A large and rich literature exists on the models of pricing financial assets. Their goals are usually (i) to reproduce the price of quoted financial assets (e.g. the prices of bonds, which gives the possibility to evaluate new over-the-counter (OTC) financial instruments, such as structured notes, path dependent options, exotic interest-rate instruments); or (ii) to make price projections and thus provide evaluations of the risk exposure of assets portfolios. Using models, it is possible to choose a risk-management policy and to decide hedging ratios. More important than the theoretical evaluation of financial assets, determining the hedging ratios is often the essential goal of these models.

Two main approaches exist for modeling the interest rates structure and its dynamics. *The first approach is based on a priori hypotheses on the functional form of the drift and on the variance of the stochastic process used to describe the dynamic of the independent variables.*

The addition of equilibrium conditions on the markets [2.01] leads to a partial differential equation (PDE). The PDE depends on the parameters of the model and on the market price of risk. Different methods have been used to estimate the latter. The first method is to estimate the model parameters by fitting historical time series. Once the parameters are estimated, the price of risk can be found by adjusting the discount function obtained by the current discount bonds prices (in which case, the fitting of historical data can be quite good but the risk price found is frequently unacceptable). The second way is based on the simultaneous estimation of the model parameters and of the price of risk from the zero-coupon bond yield at different time.

Another solution consists to add new hypotheses regarding the price of risk [2.02]. The advantage of this hypothesis is that the price of risk is explicitly estimated but it leads to imperfect fitting of the observed prices (the difference between theoretical and observed prices becomes significant).

Since the general equilibrium model leads to cross-sectional restrictions on the interest rates structure, one can use the general method of moments (GMM) to estimate the parameters and to test the restrictions (Hansen, 1982; Longstaff and Schwartz, 1992; Chan et al. 1992). However, this method raises practical problems: for a high number of parameters (Longstaff and Schwartz, 1992) the model may fit very well the observed term structures, but the estimates correspond to a local minimum. The result may be an erroneous estimation of the sensitivity of prices to the parameters. The determination of hedging ratios on this basis remains difficult.

When the number of parameters is small (Cox et al.,1985b) the fitting of the observed prices is poor, especially for specific term structures. Finally, we note that the PDE rarely accept closed form solutions (such a solution is possible when the independent variables have simple form). If no closed form solution exists, the only way is through numerical methods such as Monte-Carlo.

The second approach assumes that the initial interest rates structure is known and that the interest rates dynamic is exogenous. For example, the Ho and Lee (1986) model is based on the hypotheses that the initial spot interest rates structure is given and that the dynamic of the interest rate is described by a binomial process. As it is a one-factor model, the shocks on the different maturity are perfectly correlated. Consequently, it appears to be possible to cover a 30-year position by a short-term treasury note! More recently, Heath et al. (1992) have developed a methodology which, assuming a given initial forward interest structure, leads to a family of processes that respects the arbitrage-free condition. The use of these models goes through a fitting step that can be done using either historical observed data or implied volatility of derivatives.

Both approaches require a fitting procedure on historical data. The non-parametric approach developed in this chapter is different. It does not assume a priori hypotheses on the functional form of the process generating the interest rates structure and on the form of the distribution that characterizes the dynamic of the random variables observed. Using a historical data set, a SOM algorithm is used to approximate both the distribution of the interest rates structure and its deformations over time (the shocks on the interest rates structures). On this basis, a Monte-Carlo simulation gives long-term interest rates structure evolution.

We briefly introduce the SOM algorithm to emphasize its useful features in the context of our application. We also introduce the Monte-Carlo simulation and the GMM algorithm. These are the methodological building blocks of this chapter. The third section is devoted to the presentation of the proposed approach. Its application to a real data set is described in Section 2.4. The fifth section gives a validation procedure based on the Cox et al. (1985b) model. Finally, in Section 2.6 the approach is extended to the value-at-risk concept.

2.2 Building Blocks

2.2.1 The SOM Algorithm

The SOM algorithm is a well-known unsupervised learning algorithm which produces a map composed of a fixed number of units (Kohonen, 1982, 1995; Cottrell and Fort, 1987; Cottrell et al., 1994). After learning, each unit represents a group of individuals with similar features. The correspondence between the individuals and the units (more or less) respects the input space topology: individuals with similar features correspond to the same unit or to neighboring units. The final map is said to be a self-organized map which preserves the topology of the input space.

While the asymptotic properties of this algorithm are not rigorously proved, some of its theoretical properties have been demonstrated. One of them is of particular interest: it is the density approximation property. Pagès (1993) shows that a SOM algorithm terminating with a zero neighbor at the end of the learning converges; this is equivalent to the convergence of a classical vector quantization technique or competitive learning. The author shows that the units after vector quantization are a good discrete skeleton for reconstructing the initial density (provided that each unit is weighted by the probability estimated by the frequency of its Voronoi region). Provided that units are adequately weighted, this result shows that it is possible to reconstruct the initial data, and this result is exact when the number of units goes on to infinity.

This remarkable property is true because the SOM algorithm is nothing else than a usual competitive quantization. The classes that we obtain are topologically ordered, they can be represented in a convenient way; and they can be easily grouped if necessary. Moreover, the convergence is accelerated by the non-zero neighbor phases [2.03]. This justifies the choice of the SOM algorithm to quantify the distributions of interest rates structures and interest rate shocks and largely explains the compatibility of the generated scenarios with the historical data.

2.2.2 Monte-Carlo Simulation

The Monte-Carlo simulation has largely been used to estimate the value of derivatives when no analytical solution exists. Boyle (1977) introduced this method for the first time to estimate option's prices. The method consists in generating a great number of paths with a discrete step. Advances forward in time depend on a drift and volatility which are a priori defined.

The procedure for asset valuation can be described as follows: first, simulate a large number of paths for the explicative random variable, second, calculate the cash flows corresponding to the state of this variable and, finally, compute the present value of the calculated cash flows. Theoretically, the estimation error of the mean can be set arbitrarily. As the precision level is proportional to the square root of the number of paths, an increased degree of precision quickly leads to a high increase of the calculation load. To solve this problem, various techniques have been used which reduce the calculation time by one order [2.04].

2.2.3 The General Method of Moments

In the validation procedure (see Section 2.5), we will use the Cox et al. (1985b) model to generate theoretical paths (CIR paths) for interest rates structure. The CIR paths will be used as historical data that are necessary to calibrate the non-parametric approach. Once the calibration procedure is achieved, we will generate several simulated paths that should have the same properties as the CIR paths.

To test the properties, we will use the General Method of Moments (GMM) of Hansen (1982). With GMM, we estimate the parameters of the process underlying the simulated paths and then we compare these estimates with the initial parameters used to generate the CIR paths.

The GMM approach can be used when the economic model respects the constraint that the mean product of the error term and the observed variable must be equal to zero. In practice, the mean is replaced by the value calculated on the data set. GMM estimates the true value of the parameters by forming well-chosen linear combinations of the orthogonal constraints. This method requires no hypothesis about the form of the error distribution and the GMM estimations remains consistent even if the residuals appear to be serially correlated in time (or exhibit heteroscedasticity). Only the stationarity of the generating process and the ergodicity of the explanatory (the short rate in the CIR model) variables are needed to insure the asymptotic convergence of the estimator.

In other words, GMM relies on two conditions that can be intuitively formulated as such:

- if the model explains the observed rates, the empirical residuals mean must be zero;
- if the model explains the observed rates, the empirical covariance between the residuals and the short rate (the explanatory variable in the CIR model) must also be zero.

Hansen gives an objective function that allows the estimation of the parameters by GMM.

2.3 Simulating Future Behavior using Historical Information

To simulate the future behavior of a process using historical data, we start from a matrix composed of process observations. The process can be characterized by one or several observed variables. The initial matrix, denoted D, is of size $[T'\ P]$, where T is the number of observations and P is the number of observed variables [2.05].

The first step of our approach is to choose a lagging order. The initial data matrix, D, is then modified to incorporate, for each row, the vector of the observed variables as well as the past realizations. The new data matrix, LD, is $[(T-\lambda)\times(P\times\lambda)]$, where l is the lagging order. Rows of LD are denoted $c_t = \{c_{t,1},\dots,c_{t,LP}\}$, where t is the time index and $LP = (P\times\lambda)$.

The LD matrix is then decomposed into a number of homogeneous clusters, using the SOM algorithm (a one-dimensional map is used). Each unit of the map provides for a number of winning individuals. For each unit, the mean profile of the attached individuals is calculated. The choice of the number of clusters will depend on the features of the analyzed process. The homogeneity of the clusters is measured, for example, by the Fisher statistics or one of its multidimensional extensions obtained after learning.

For each row x_t of the LD matrix, the associated deformation (or shock) is then computed. It is denoted y_t and is obtained by the following calculation: $y_t = c_{t+\tau} - c_t$, where t is a time delay. On this basis, for each cluster of the LD matrix $(T-\lambda, LP)$, a P_i matrix whose rows are the y_t corresponding to the x_t in cluster i, is formed.

For each P_i matrix can be decomposition into a number of homogeneous using the SOM algorithm. The mean profiles of the formed clusters are then determined.

The last step to characterize the analyzed process is to calculate the empirical frequencies of $\overline{y}_{i,j}$ conditionally to $\overline{x}_i$.

The simulation procedure takes the following form:

1. choose a starting point x_t (for example, one row of the LD matrix);
2. determine the winning $\overline{x}_i = \mathrm{ArgMin}\|x_t - \overline{x}_i\|$,
3. randomly pick according to the conditional distribution $P\left(\overline{y}_{i,j}\middle|\overline{x}_i\right)$;
4. compute $x_{t+1} = x_t + y$;
5. iterate the procedure for simulating the dynamics of the process on a specific time horizon.

For stochastic processes, the procedure will be iterated and the results will be averaged.

2.4 Application

2.4.1 Data Set

To simulate interest rates structures evolution on a long-term horizon, we used data from the US bonds market. Our data are daily interest rates structures with

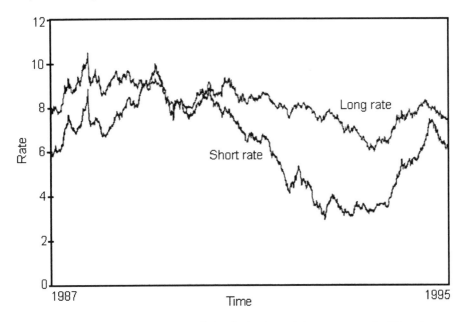

Figure 2.1. Short-rate and long-rate evolution during the period 1987–1995 on the US market.

maturity from 1 to 15 years. The interest rate for each maturity has been calculated by JP Morgan from the prices of US T-Bills and T-Bonds. The sample covers the period from 1/5/1987 to 5/10/1995, altogether 2088 entries. From these data, we calculate the differences between the observed term structure at time t (given that we have only 15 rates corresponding to maturity ranging from 1 year to 15 years) and time t-10 working days (time delay recommended by the Basle Committee on Banking Supervision). Figure 2.1 shows the evolution of the short-rate (1 year) and the long-rate (15 years) during the period 1987–1995. The results presented in this chapter have been confirmed on different data sets.

2.4.2 Results

2.4.2.1 The Classification of Interest Rates Structures

To classify the interest rates structures, we have used a one-dimensional map formed by nine units [2.06]. The mean profiles obtained are presented in Figure 2.2. From the nine mean profiles, only the shape of unit 1 is really specific (with a short rate superior to the long rate). There are 169 structures belonging to this category, that is to say 8% of the total data. The other eight mean profiles have a more or less comparable shape and are separated by the level of the short rate [2.07].

Note that the exclusive results of a PCA [2.08] on the normalized data would have probably missed the importance of the 169 structures attached to the first unit.

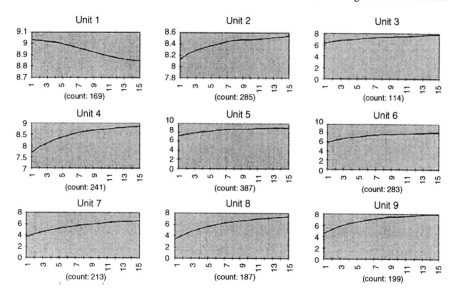

Figure 2.2. Mean profiles of the clustered interest rate structures, using daily data coming from the US market (1987–1995).

Indeed, considering that the weight of the first factor is very high (97%) and that its correlation with the level is near perfect, we would have thought that the only important variable is the level of the short rate. While it is clear that it explains a very important part of the total variance, the spread and the curvature must also be considered.

Finally, we note that these results are compatible with the widely used study of Litterman and Scheinkman (1988). During the period from January 1984 through June 1988, the authors show that on average 97% of the variance of the interest rates is explained by a three-factor model for weekly observations. The first factor is called the level factor and is highly correlated with the level of the short rate. The second factor, the steepness, corresponds to a variation of the slope (in other words, of the spread) and the third factor is called the curvature. We found in our study that, for daily observations, in 50% of the total number of cases the main deformation is a variation of the slope.

Table 2.1. Principal components analysis on interest rate structure, using daily data from the US market (1987-1995)

Explained variance			Correlations		
			Factor 1	Factor 2	Spread
Factor 1	0.97%				
Factor 2	0.03%	Level	0.937	-0.344	-0.924
Factor 3	0.00%	Spread	-0.734	0.677	
Factor 4	0.00%				

2.4.2.2 The Classification of Interest Rates Shocks

The classification of interest rates shocks has been realized using a 30-unit one-dimensional SOM map (for the same reason as quoted above). Figure 2.3 highlights the main results obtained at this stage. It presents the mean profiles of the interest rate shocks attached to each unit and graphically shows that the produced classification may be explained by a variation of the spread – the difference between the long rate and the short rate – (for example, the mean profile of Unit 1) by a change in the curvature (for example, the mean profile of Unit 30) and by the level of the short rate (for example, the mean profile of Unit 22).

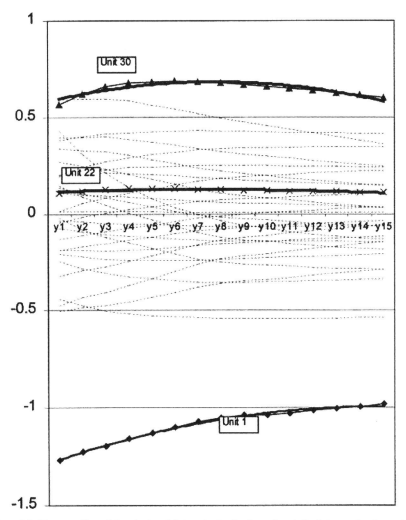

Figure 2.3. Mean profiles of the clustered interest rate shocks, using daily data coming from the US market (1987–1995).

2.4.2.3 The Conditional Distributions

A large number of term structure models are based on specific assumptions about the stochastic process determining the observed variables, especially the instantaneous interest rate. Various empirical studies conclude that some classes of models seems to outperform other models. This class of models considers the mean-reverting feature of the interest rates. The changes in the variance appear to be dependent on the level.

As underlined in the introduction, the method developed in this chapter does not impose an a priori analytical form of the process governing the instantaneous interest rate. But if there is a mean-reverting feature or a link between the interest rate level and the variance, our approach will consider it. In fact, as we characterized the relation between the class of interest rate shocks and the initial structure (the state of the process), we consider the magnitude of the shocks (variance) with respect to the initial structure. The probability of an upper interest rate shock will be lower when the level of the initial structure is high and vice versa (which is intuitively the natural consequence of a mean-reverting property). Moreover, as we do not impose the processes governing the observed variables, we hope to reproduce the statistical properties of the historical evolution of the interest rates structure.

The analysis of the relation between the initial interest rates structure and the shocks that apply to it has been conducted on the basis of the results of the clustering of the interest rates structures and interest rate movements. To understand this procedure, we have to remember that each interest rate shock is related by its date to a specific initial interest rates structure. In other words, a specific movement of the interest rates structure is obtained by the difference between the state of the interest rates structure at a specific date and the state of the interest rates structure ten days later. Therefore, it is possible to identify the class of shocks associated with each of the nine interest rates structure classes and to calculate the nine corresponding conditional frequency distributions (see Table 2.2).

We tested the statistical independence between the nine empirical conditional distributions of shocks and the global population of shocks using a χ^2 test. The results are shown in Table 2.3. The conditional distributions of interest rate shocks are statistically different from the distribution of the global population of shocks, with a very high level of confidence. The existence of a relation between shocks and initial interest rates structure is empirically confirmed.

2.4.2.4 Simulating Paths on a Long-term Horizon

Using these empirical conditional distributions of the evolution of the frequencies, we use the Monte-Carlo procedure described in Section 2.3 to simulate the interest rates structure. In the case of interest rates, the procedure becomes:

1.   draw randomly an initial interest rates structure;
2.   determine the number of the Kohonen class of this interest rates structure;

Table 2.2. Nine frequencies distributions of interest rate shocks conditional on the interest rate structure classes

Classes of shocks	Classes of structures								
	1	2	3	4	5	6	7	8	9
Cluster 1	0	0	0	0	0.02	0	0	0	0
Cluster 2	0.02	0	0	0	0.02	0	0	0	0
Cluster 3	0.07	0	0	0.01	0	0.01	0	0.01	0.02
Cluster 4	0.02	0	0.01	0.04	0.02	0	0.02	0.02	0.02
Cluster 5	0.01	0.02	0.02	0.03	0.04	0	0	0.04	0
Cluster 6	0.06	0.03	0.01	0.03	0.03	0.04	0	0.03	0.04
Cluster 7	0.01	0	0	0.01	0.02	0	0.02	0.03	0
Cluster 8	0.07	0	0	0	0	0.01	0	0	0.06
Cluster 9	0.08	0.09	0.08	0.06	0.03	0.07	0.09	0.02	0.06
Cluster 10	0.06	0.09	0.13	0.09	0.09	0.06	0.07	0.05	0.03
Cluster 11	0.01	0.07	0.08	0.05	0.07	0.09	0.04	0.06	0.02
Cluster 12	0.07	0.03	0.04	0.09	0.04	0.01	0.06	0.07	0.06
Cluster 13	0.03	0.06	0.02	0.08	0.03	0.05	0	0.05	0.05
Cluster 14	0.02	0.03	0	0.03	0.03	0	0.01	0.07	0.02
Cluster 15	0.06	0.02	0.03	0.02	0	0.04	0.01	0.06	0.04
Cluster 16	0	0.01	0	0	0.03	0	0	0.01	0
Cluster 17	0.03	0.05	0.04	0.03	0.06	0.09	0.08	0.06	0.03
Cluster 18	0.03	0.05	0.11	0.07	0.04	0.08	0.15	0.04	0.05
Cluster 19	0.08	0.05	0.06	0.05	0.02	0.05	0.05	0.04	0.06
Cluster 20	0.05	0.04	0.05	0.07	0.03	0.09	0.04	0.06	0.09
Cluster 21	0.01	0.02	0.01	0.04	0.05	0.02	0.02	0.02	0.02
Cluster 22	0.05	0.04	0.01	0	0.01	0.04	0.02	0.05	0.06
Cluster 23	0.05	0.07	0.04	0.04	0.07	0.07	0.08	0.05	0.08
Cluster 24	0.01	0.04	0.01	0.03	0.08	0.03	0.04	0.05	0.05
Cluster 25	0.04	0.05	0.08	0.06	0.03	0.08	0.05	0.04	0.07
Cluster 26	0.02	0.04	0	0.01	0.02	0.03	0.04	0.03	0.04
Cluster 27	0.03	0.04	0.05	0.03	0.05	0.02	0.04	0.02	0
Cluster 28	0.01	0.04	0.03	0.03	0.03	0.01	0.05	0	0.02
Cluster 29	0	0.02	0.03	0	0.03	0.01	0.02	0.01	0.01
Cluster 30	0	0	0.06	0	0.01	0	0	0.01	0

Table 2.3. χ^2 tests between each of the nine conditional distributions of shocks and the global population of shocks

Initial interest rate classes	Independence test
1	140
2	42.04
3	85.48
4	54.58
5	151.07
6	244.39
7	55.38
8	73.37
9	59.87

3. draw randomly a shock according to the conditional distribution of frequencies of the interest rate shocks;

4. apply the shock to the interest rates structure;

5. repeat the procedure 125 times to construct an interest rates structure evolution on a 5-year horizon (125 times the 10 days covered by the interest rate shock);

6. for each simulation, repeat the procedure 1000 times to build the distribution of probability of interest rates structures, starting from the same initial interest rates structure.

Figures 2.4 and 2.5 respectively show the distribution of the short rate and the long rate for three simulations. The first two have been realized using the same interest rate initial shape (for which Unit 6 is the winning one). The third one has been done using an initial interest rates structure attached to Unit 1 (the only inverted interest rates structure mean profile).

Based on these figures, we see that the procedure is stable and that, on a 5-year basis, the initial interest rates structure mainly influences the short rate level. For all simulations, the level of the short rate and the long rate are compatible with the historical one. Figure 2.5 presents five interest rates structures obtained by the first simulation, drawn among 1000 interest rates structures. Figure 2.6 presents one path of the short and the long rate over 5 years as produced by the first simulation. These clearly represent possible interest rates paths. We should also mention that in all simulations and at all steps all forward interest rates are positive.

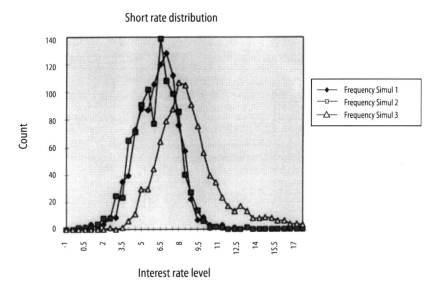

Figure 2.4. The short-rate distributions produced by simulations 1 and 2 (starting from the same initial interest rate structure) highlight the stability of the simulation procedure.

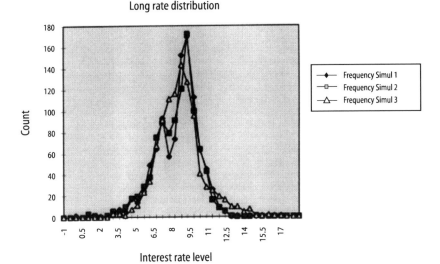

Figure 2.5. The long-rate distributions produced by simulations 1 and 2 (starting from the same initial interest rate structure) highlight the stability of the simulation procedure.

2.5 Validation

Among the many open questions that remain about the approach that we propose here, one of the most important is the notion of *compatibility* of the simulated paths with the historical data set used. By compatibility we mean that the simulated paths will, on average, exhibit the same statistical properties as the process underlying the historical data set.

To test this property, we propose the following approach. A *theoretical* data set of interest rates structures will be generated from the well-known CIR interest rates structure model. From there, we will use the approach described in Section 2.3 to produce simulated paths. On each simulated path, we will estimate, using GMM, the parameters of the process. Then, we will verify if the parameters are the same as those that were used to generate the theoretical data.

Figure 2.7 shows a path of 2000 steps for the short rate via this procedure. The theoretical time series of the short rate are used to calculate, at each step, the interest rate for the maturity going from 1 year to 15 years. The short rate and the spot interest rate ranging from 1 year to 15 years represent the historical data set. The next step of the validation procedure is to use the approach depicted in Section 2.3. First, the clustering of 2000 interest rates structures is realized. The one-day shocks of the interest rates structures are computed. The clustering of the shocks associated to each class of interest rates structure is then realized. Finally, the Monte-Carlo approach is applied to simulate possible future paths. Using these paths, the GMM algorithm estimates the parameters of the CIR interest rates

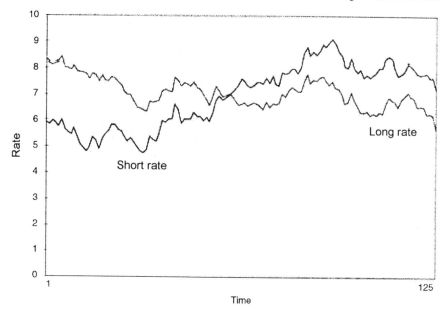

Figure 2.6. One trajectory of the short and long rate over 5 years, chosen among the 100 produced by simulation 1.

structure model. The estimates are compared to the initial values that are extract from Dahlquist (1996).

Table 2.4 presents a first set of results. The rows give the mean estimation of parameters a, b and s obtained after five simulated paths of 2000 steps. The results show that estimation of the parameters is very good.

2.6 Extension to Value-at-Risk

The classical approach to model risk is to assume that the asset's returns are normally distributed or that the prices are lognormal. The classical tradeoff between risk and return is represented in a two-dimensional space: mean (return) and variance (risk). If the returns have normal multivariate distribution, the risk of a portfolio only depends on the matrix variance-covariance of the returns.

Value-at-Risk (VaR) is an estimate of the maximum loss to be expected over a given period a certain percentage of the time. Risk in this case is measured by the maximum potential loss that can be estimated by historical or Monte-Carlo

Table 2.4. Mean estimation of the parameters obtained with the GMM algorithm on five simulated paths of 2000 steps in time

Parameter	Input	Estimate	Error	t-statistic
b	0.7037	0.7037	5.74E-09	1.26E+08
s^2	0.0144	0.144	2.70E-09	5.34E+06
a	0.0902	0.0902	6.10E-10	1.49E+08

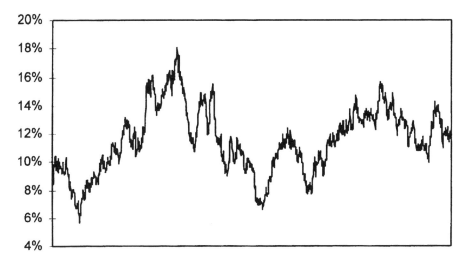

Figure 2.7. A theoretical path of the short-rate obtained using the CIR model with Dahlquist parameters set for 2000 steps in time.

simulations. Historical simulations give the maximum loss over an observed period of time. The limits of the historical approach are that the results depend on a past period while the Monte-Carlo simulations depend on the matrix variance-covariance.

A recent study (Beder, 1995) shows that the magnitude of discrepancy among these methods deviates by more than 14 times for the same portfolio. Moreover, the VaR results that are calculated by an historical simulation depend on the behavior of the return during the period. For example, during a period of decreasing interest rates, risk may be underestimated.

Historical simulations face all the problems encountered by the non-experimental science, while Monte-Carlo simulations are well adapted for very short periods of time (from 1 day to 1 month). For example, the conceptual framework developed in RiskMetrics (JP Morgan) assumes that historical returns are generated from a multivariate normal distribution. From this assumption, future price paths can be simulated by the Cholesky decomposition. This procedure is adapted to a short-term horizon but fails when applied to the long term. The generated scenarios tend to be explosive because the resulting dynamics of the interest rates structure do not have mean reverting properties.

The generation of the long-term scenario still remains a problem hard to solve. The approach that we propose in this chapter solves this specific problem. As shown in Section 2.4, we are able to produce simulated paths on a long-term horizon (5 years or more), which are non-explosive, producing only positive forward rates and leading, through the Monte-Carlo procedure, to the construction of distribution of rates on each maturity on any desired time horizon. Moreover, as we begin to show in Section 2.5, the simulated paths seem to capture the dynamic of the process that has generated the historical data set.

Application of this procedure to VaR is quite straightforward. On the basis of the simulated paths of interest rates structure, we can evaluate path-dependent assets' prices over time. This will lead to distribution of prices, shedding light on the level of risk exposure, on a selected time horizon, borne by the financial institutions which own positions on those assets.

2.7 Conclusions

In this chapter, we have proposed a procedure to generate a long-term path of interest rates structures. The procedure involves a general algorithm which captures the properties of a process dynamics observed through a set of historical data. This algorithm, described in Section 2.3, uses two SOM maps to form clusters of similar vectors of data and of similar deformations in time. The empirical distribution of frequency of deformations conditional to the initial vector is then computed. We used a Monte-Carlo simulation to generate paths of the observed process. Applying this procedure to interest rates structures, we showed that the procedure does not generate explosive paths, even on a long-term horizon.

To test the accuracy of the procedure we generated a set of vectors of interest rates from a theoretical model. The data was generated by the well-known Cox–Ingersoll–Ross interest rate model. We also used the General Method of Moment to verify if the simulated paths have the same properties as the theoretical path. The results are encouraging, although they are not complete enough to assure that the proposed method reproduces simulated paths with the same properties as the historical path.

In the future we will focus on improving our validation procedure; we will extend our work to the two-factor model and will investigate the possible applications of this simulation procedure to (a) estimate Value-at-Risk for interest rates contingent claims such as mortgage securities; (b) reproduce the observed prices on the market; and (c) calculate the hedge ratio.

3 Picking Mutual Funds with Self-Organizing Maps

Guido Deboeck

Abstract

The self-organizing map approach can be used to translate multi-dimensional mutual fund data into simple two-dimensional maps. These maps provide a significant improvement over the information that is traditionally published on mutual funds. They create a better basis for portfolio selection, for comparison of the performance of mutual funds and for the creation of benchmarks.[1] Using data published by Morningstar, we use self-organizing maps to demonstrate clear distinctions and patterns among the best rated mutual funds. As inputs we used the performance of the mutual funds, the Morningstar risk measurements, the tenure of investment managers, and the various expense ratios. SOMs simplify classification of funds; can be used for decision-support; and to provide classifications that have more meaning than simple sorted lists based on multiple criteria.

3.1 Exploratory Data Analysis

As pointed out in the introduction, exploratory data analysis and data mining are used for knowledge discovery in large databases (Kaski, 1997). The self-organizing map method is used to search for patterns among seemingly similar mutual funds. The main reason for using SOM are:

1. it is a numerical instead of a symbolic data mining method;
2. it is a non-parametric method, meaning that no a priori assumptions about the distribution of the data need to be made;
3. it is a method that can detect unexpected structures or patterns by learning without supervision.

---

[1] We thank Professor Mark Embrechts, Associate Professor, Department of Decision Sciences and Engineering Systems at Rensselaer Polytechnic Institute and Dr. Samuel Kaski of the Helsinki University of Technology for their comments on earlier drafts; and we thank in particular Antti Vainonen for processing the SOM maps in this paper.

SOM is a neural network technique that learns without supervision. In contrast to other neural network techniques – several of these techniques were discussed in Deboeck (1994a) – which require that one or more outputs are specified in conjunction with one or more inputs in order to find patterns or relationships between data, SOM reduces multi-dimensional data to a two- or three-dimensional maps or grids of neurons. The SOM algorithm is based on unsupervised, competitive learning. Map units, or neurons, usually form a two-dimensional grid, and thus the mapping is from one high-dimensional space onto a plane (or cube). The topology of the data is preserved, which means that a SOM groups similar input data vectors on neurons or grid points. Points that are near each other in the input space are mapped to nearby map units in the SOM. The SOM can thus serve as a cluster-analyzing tool for high-dimensional data.

A SOM has also the capability to *generalize*. Generalization capability means that the network can recognize or characterize inputs it has never encountered before.

3.2 Morningstar Mutual Fund Database

Morningstar is a privately owned company founded in 1984 to provide investors with useful information for making intelligent, informed investment decisions. Morningstar publishes *Morningstar Mutual Funds*, an in-depth analytical magazine on mutual funds *Morningstar Investor,* is a monthly compendium of 450 open-end and 50 closed-end funds selected from a universe of more than 6000 mutual funds. The 500 represent, according to Morningstar, the *most consistently successful funds in the mutual fund industry today.*

The data sets used were derived from this selection of 500 funds. Morningstar Ascent (Morningstar, 1997) was used to select and group mutual funds. Ascent's key features include filtering, custom tailored reports, detailed individual-fund summary pages, graphic displays, and portfolio monitoring. It does not provide any exploratory data analysis nor does it allow automated data mining for finding patterns among seemingly similar mutual funds. In this chapter we show how exploratory data analysis can be used for improving selection and construction of a mutual fund portfolio. This could yield yet another revision of the Morningstar rating system (Regnier, 1996).

3.3 A Simple Binary Example

In selecting mutual funds, investors often look at several criteria: the return achieved over one, three or five years, the risk or volatility of the returns vis-à-vis a benchmark, the size of the fund, the turnover rate, the expense ratio or the management of the fund, and the tenure of the management of a fund.

Performance and risk related measurements of funds are abundant. It is easy to find statistics on *performance,* including the performance over the last quarter, the last twelve months, since the beginning of the year, the last three years, five years, ten years, or in some cases even fifteen years.

Risk on the other hand is often measured as the volatility of returns, and measured as the standard deviation of returns. Another indicator that helps to assess risks, also provided by Morningstar, is the decile rank of the performance of a fund in a bear market.

The *size of a fund* or the amount of assets under management is often taken as a proxy of maturity of a fund: young startup funds have small amounts of assets under management; longer established funds can have much larger assets under management. The *turnover rate* or percentage of a portfolio that is changed during a year provides an indication of how actively a fund is managed (turnover rates of 100% or more often indicate very active management).

An important criterion investors often look for is the *tenure of the management* of the fund: funds that have management with longer tenure are assumed to perform better than those that have very short tenured management. Of course, as a good friend of mine recently pointed out, the age of the manager may be more important because some funds are managed by managers who have very long tenure, but also are in their 70's. Unfortunately, much information on the background and experience of fund managers still remains scarce.

All of the these (and potentially) other factors are not necessary quantifiable. Methods that can provide equal weight to various attributes, including those that cannot be turned into discrete or continuous values, are desirable. Attributes that attain quantitative properties such as "present" or "not present" can be represented by a binary value, i.e. 1 or 0. The similarity between binary attributes can then be defined in terms of the number of attributes common to both sets.

To illustrate this we present first the results of a simulation based on the data in Table 3.1. Eighteen mutual funds were selected from the Morningstar mutual fund list of best performing international equity funds. The presence or absence of 20 different attributes of these mutual funds are given. The first four attributes sub-classify the international equity mutual funds according to their objectives: funds mainly investing in foreign equities, world equities, open-ended funds investing in emerging markets, and closed-end funds. The next three attributes indicate whether the three-year annualized return of a fund was higher, lower or about equivalent to the three-year annualized return of the EAFE (European and Far East) index. The next attribute uses the Morningstar rating of the fund to indicate whether it had a category 3, 4, or 5 rating. The following attribute records the number of years of management tenure (equal or less than 2 years, 3–4 years, about 5 years, 10 years or more). The final attribute concerns the expense ratio of the fund (less than 1%, between 1% and 1.25%, and 1.25% or higher).

Each column is an input vector for the SOM. The fund name and the fund symbol do not belong to the vector but specifies the label of the fund in the calibration of the self-organizing map [3.01]. The data set in Table 3.1 was presented iteratively and in a random order to SOMs of different sizes. The initial model vectors between the neurons were chosen to be small random values, i.e. no prior order was imposed. However, after 1000 presentations, each neuron became more or less responsive to one of the occurring attribute combinations and simultaneously to one of the 18 mutual funds. Thus we obtained the map shown in Figure 3.1. The labels on the map refer to mutual fund symbols provided in Table 3.2.

Table 3.1. Sample binary data matrix of selected mutual funds

			1	2	3	4	5	6	7	8	9	10	11	12	13	14	15	16	17	18
Attributes		Fund name	Harbor International	T. Row Price International	Scudder International	Templeton Foreign	Vanguard International Growth	Fidelity World	Janus World	Putham Growth	Scudder Global	Templeton Global Small Companies	Templeton Growth	Merrill Lynch Dev Cap	Templeton Developing Markets	Emerging Mkts Telecommunications	Latin Am Equity	Morgan Stanley Africa	Morgan Stanley Emerg Mkts	Templeton Emerging Markets
		Fund symbol	HBI	TRP	SCI	TFO	VIG	FWO	JWO	PUG	SCG	TGS	TGI	MLD	TDM	EMT	LAC	AFR	MSE	TEM
1	Type IS	foreign	1	1	1	1	1	0	0	0	0	0	0	0	0	0	0	0	0	0
2		world equity	0	0	0	0	0	1	1	1	1	1	1	0	0	0	0	0	0	0
3		emerging market	0	0	0	0	0	0	0	0	0	0	0	1	1	1	1	1	1	1
4		closed-end	0	0	0	0	0	0	0	0	0	0	0	0	0	1	1	1	1	1
5	Total Return	= > EAFE	1	0	0	1	1	1	1	1	1	1	1	0	0	0	0	0	0	0
6	- 3 yr annualized	about EAFE	0	1	1	0	0	0	0	0	0	0	0	0	0	0	0	0	0	0
7		< = EAFE	0	0	0	0	0	0	0	0	0	0	0	1	1	1	1	1	1	1
8	MorningStar	average 3	0	0	1	0	0	1	0	1	1	0	0	0	0	1	1	0	1	0
9	Category Rating	high 4	0	1	0	1	1	0	0	0	0	1	1	1	0	0	0	0	0	1
10		highest 5	1	0	0	0	0	0	1	0	0	0	0	0	1	0	0	0	0	0
11	Risk	low	0	0	0	1	0	1	0	0	0	0	1	0	0	0	0	0	0	0
12		medium	1	1	1	0	1	0	1	1	1	1	0	0	0	0	0	0	0	1
13		high	0	0	0	0	0	0	0	0	0	0	0	1	1	1	1	1	1	0
14	Management	< = 2 years	0	0	0	0	0	0	0	0	0	1	0	0	0	0	0	0	0	0
15	Tenure	3–4 years	0	0	0	0	0	0	1	0	0	0	0	0	0	0	1	0	0	0
16		5 years +	0	0	0	0	0	1	0	0	0	0	0	1	1	1	0	0	0	0
17		10 years	1	0	0	1	1	0	0	1	0	0	1	1	0	0	0	0	0	1
18	Expense Ratio	low < 1.0	0	1	0	0	1	0	0	0	0	0	1	0	0	0	0	0	0	0
19		medium 1–1.25	1	0	1	1	0	1	1	1	0	1	0	0	0	0	0	0	0	0
20		high 1.25+	0	0	0	0	0	0	0	0	1	0	0	1	1	1	1	1	1	1

Source: MorningStar Year-end Wrap January 1997

The first example is a SOM of 1 by 10 neurons. Figure 3.1 shows that several mutual funds cluster. For example Fidelity World (FWO), Janus World (JWO), Scudder Global (SCG), and Templeton Global Small Companies (TGS) cluster on one neuron; Emerging Markets Telecommunications (EMT), Latin American Equity (LAC), Morgan Stanley Africa (AFR), and Morgan Stanley Emerging Markets (MSE) cluster on another neuron. There are three other neurons that contain two mutual funds.

Since input vectors that are similar cluster on the same or nearby neurons, the groupings shown in Figure 3.1 indicate similarity of input attributes between the funds that are on the same or nearby neurons. To verify this we lookup in Table 3.1 the first group listed and compare it with the second major group. The first group of funds are more diversified while the second group contains funds that are more focused on specific sectors or geographical regions of the world. EMT, for example, is a fund that only invests in telecommunication companies; LAC, AFR

Figure 3.1. SOM of binary sample data matrix, using map size 10 by 1, where the initial training length was 1000 cycles, the alpha or learning rate was 0.05 with a starting radius of 5; the second phase training was for 10,000 cycles, with learning rate 0.02 and radius of 1.

Table 3.2. Key to mutual fund names and symbols

Fund name	Symbol	Fund name	Symbol
Harbor International	HBI	Templeton Global Small Companies	TGS
T. Row Price International	TRP	Templeton Growth	TGI
Scudder International	SCI	Merrill Lynch Dev Cap	MLD
Templeton Foreign	TFO	Templeton Developing Markets	TDM
Vanguard International Growth	VIG	Emerging Mkts Telecommunications	EMT
Fidelity World	FWO	Latin Am Equity	LAC
Janus World	JWO	Morgan Stanley Africa	AFR
Putham Growth	PUG	Morgan Stanley Emerg Mkts	MSE
Scudder Global	SCG	Templeton Emerging Markets	TEM

and MSE are funds that specialize on investments in Latin America, Africa or emerging markets, respectively.

The gray shading of the neurons on this map indicates that those neurons with similar shading contain mutual funds that are closer together, while neurons that are darkly shaded create gaps between groups. Thus Figure 3.1 can be interpreted to mean that these 18 mutual funds fall into five major groups. These are:

- group I with FWO, JWO, SCG, TGS as well as Putham Growth (PUG);
- group II with Harbor International (HBI), Scudder International (SCI) and T. Row Price International (TRP);
- group III with Vanguard International Growth (VIG), Templeton Foreign (TFO) and Templeton Growth (TGI);
- group IV includes Merrill Lynch Development Cap (MLD) and Templeton Emerging Markets (TEM);
- group V clusters specialized sectorial and regional-biased funds (EMT, LAC, AFR, MSE).

Our first example was a one-dimensional map where we arbitrarily determined upfront that we wanted to see how 18 mutual funds would cluster if ten model vectors were provided. The result was a clustering of 18 funds into five major groups (because mutual funds with light shades in between them are similar, whereas funds with dark shades in between them are relatively dissimilar). In the second example we created a two-dimensional map by providing a grid of model vectors. We arbitrarily defined upfront a grid of 6 by 4 neurons. After an initial 10,000 presentations (starting with initial learning rate of 0.5 that was decreased linearly as the data presentations progressed, and a secondary learning phase of 100,000 presentations starting with a learning rate or 0.2) we obtained the map shown in Figure 3.2.

Twenty-four model vectors were provided, but only 11 attracted any mutual funds (Figure 3.2). The grayscaling of the neurons allows us to more clearly define groups or zones on the map. For example, we see a clear demarcation of the funds to the left (MLD, AFR, TEM, EMT, LAC and MSE) as compared with all other funds on the map. We also note a separation between the funds in the top two lines of neurons versus those on the bottom line of neurons. We note that TDM falls on one neuron in a corner at the bottom of the map, and that SCI is more isolated in

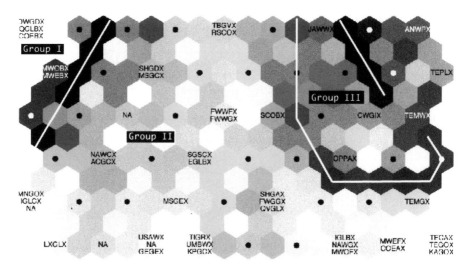

Figure 3.2. SOM of binary sample data matrix, using map size 6 by 4.

the middle of the map. Again, the grayscale grouping of funds on similar coded neurons indicates more similarity among funds falling on similar shaded neurons, while darker or more isolated neurons reflect dissimilarity of mutual funds from the rest of the group.

One conclusion drawn from these maps is that depending on the purpose of the map, the larger the map the more likely it will be that each input data vector will be attracted or characterized by a different neuron. *Large maps make good lookup tables for new data, but the generalization ability of those maps may suffer.* For example, if a new fund is established that focuses on investments in, say, sub-Saharan Africa, a good SOM map of funds investing in world stocks could be used test whether the new fund resembles more the AFR, EMT, LAC, MSE group identified above or more the TDM, TGI, JWO, PWO, TGS, VIG group in Figure 3.2.

Smaller maps provide more data compression as evidenced by Figure 3.1. However, maps that are too small may provide only a very coarse differentiation. For example, if only three model vectors were given upfront, all 18 mutual funds could be clustered onto three neurons and not provide adequate differentiation between more globally diversified funds, sector-oriented funds, or geographical-oriented funds.

A map like the one shown in Figure 3.2 may be more suitable if the goal is to formulate a portfolio of mutual funds that can represent the universe under consideration. Based on Figure 3.2 one could pick one mutual fund from each group to represent the universe of mutual funds investing in international equity funds. SOMs are excellent means to design benchmarks for comparing the performance of a (sub)class of mutual funds. Benchmarks based on SOM maps of subclasses of funds can be used as yardsticks for comparison of individual funds. These are better than the use of subclass averages, because arithmetic averages delude or bias the results towards the group with the larger number of members.

Table 3.3 Sample binary mutual fund data ordered by various clusters of a self-organizing map

Row	Col.	Fund	Fo reign	Wo rld	E M E q	Cl os ed- end	>EAFE	–EAFE	<EAFE	Mg*3	Mg*4	Mg*5	Risk low	Risk med.	Risk high	Mgmt tenure <2	Mgmt ten 3-4	Mgmt ten 5+	Mg ten 10+	Ex <1	Ex 1.25>
1	1	TRP	1	0	0	0	0	1	0	0	1	0	0	1	0	0	0	0	0	1	0
1	3	SCI	1	0	0	0	0	1	0	1	0	0	0	1	0	0	0	0	0	0	1
1	5	TGS	0	1	0	0	1	0	0	0	1	0	0	1	0	1	0	0	0	0	1
1	7	JWO	0	1	0	0	1	0	0	0	0	1	0	1	0	0	1	0	0	0	1
3	2	VIG	1	0	0	0	1	0	0	0	1	0	0	1	0	0	0	0	1	1	0
3	4	HBI	1	0	0	0	1	0	0	0	0	1	0	1	0	0	0	0	1	0	1
3	6	PUG	0	1	0	0	1	0	0	1	0	0	0	1	0	0	0	0	1	0	1
4	7	FWO	0	1	0	0	1	0	0	1	0	0	1	0	0	0	0	1	0	0	1
5	1	TGI	0	1	0	0	1	0	0	0	1	0	1	0	0	0	0	0	1	1	0
5	3	TFO	1	0	0	0	1	0	0	0	1	0	1	0	0	0	0	0	1	0	1
5	6	SCG	0	1	0	0	1	0	0	1	0	0	0	1	0	0	0	0	0	0	0
7	5	TDM	0	0	1	0	0	0	1	0	0	1	0	0	1	0	0	1	0	0	0
8	2	MLD	0	0	1	0	0	0	1	0	1	0	0	0	1	0	0	0	1	0	0
8	7	EMT	0	0	1	1	0	0	1	1	0	0	0	0	1	0	0	1	0	0	0
8	7	LAC	0	0	1	1	0	0	1	1	0	0	0	0	1	0	0	1	0	0	0
9	1	TEM	0	0	1	1	0	0	1	0	1	0	0	1	0	0	0	0	1	0	0
9	4	AFR	0	0	1	1	0	0	1	0	0	0	0	0	1	0	1	0	0	0	0
9	6	MSE	0	0	1	1	0	0	1	1	0	0	0	0	1	0	0	0	0	0	0

The above maps can also be used for feature extraction. By ordering the sample input data according to the groups described, it is possible to *extract rules* or *describe group attributes*. An example of this is provided in Table 3.3. The common features of EMT, LAC, TEM, AFR, and MSE mutual funds – which in the above SOM maps show up in the same clusters – is that they all underperformed the EAFE index over the past three years, all have expense ratios greater than 1.25%, and all are closed-end funds. None of these features is correlated with management tenure, the risk taken by the managers of these funds, or their Morningstar rating (since the basic criteria for the initial selection of 18 mutual funds investing in international equities was that they had a Morningstar rating of four or above).

We also note that TGS, JWO, VIG, HBI, PLUG, FWD, TGI and TFO are all mutual funds that have exceeded the three-year annualized EAFE index, and the managers of these funds take lower risks. The performance record of these funds is again not correlated with management tenure, the fund's expense ratio, or the Morningstar rating.

In sum, appropriated sized SOMs allow differentiation between seemingly similar mutual funds and extraction of the main features of each cluster or group of clusters. These main features may be desirable selection criteria for composing mutual fund portfolios or for the creation of benchmarks.

3.4 Mapping Mutual Funds

In the previous section we showed a simple example of the application of SOM to mutual funds. We now apply the same approach to collections of mutual funds that focus on investments in world stocks, international bonds, domestic – i.e. US-based – large capitalization's stocks, domestic small capitalization stocks, and emerging markets. All data sets for these models were extracted from Morningstar mutual fund database using the Ascent program. All data sets for these models use discrete or continuous variables rather than the binary abstractions used above. We discuss first the input data sets, then the methodology, and finally the results that emerged from each map.

3.4.1 Input Data Sets

The main variables selected from the Morningstar mutual fund database for the creation of the maps in this section are shown in Table 3.4. The main variables were management tenure; the fund's return over 12 months, three years, and five years; the Morningstar rating; the fund's decile rank in bear markets; the mean and standard deviation of returns over three years; the size of the fund as measured by the net assets under management; and the costs of participation (as expressed by front and deferred loads, and the expense ratio). Where front or deferred loads were not available for a majority of the funds the data set omitted these columns.

A typical input data set is shown in Table 3.5. All input data was scaled so that the variance of each column equaled one. This was achieved by dividing the elements by the square roots of their corresponding variances. Thus, for each element, the difference between the samples contributes equally to the summed distance measure between input sample vectors.

3.4.2 Computation of Maps

The input data sets used for the maps discussed below consisted of different selections of mutual funds. The basic criterion used for the selection of mutual funds was the investment focus of the funds as defined by Morningstar. Among the sub-

Table 3.4. Input variables and dimensions

Dimension	Main variables
Management experience	Management tenure
Return	Total return 12 months
Return	Total return annualized 3 years
Return	Total return annualized 5 years
Return	Load adjusted return 12 months
Return	Load adjusted return 5 years
Overall rating	MorningStar rating
Risk	Bear market decile rank
Return	Mean return 3 years
Risk	Std dev 3 years
Size	Net assets in $millions
Transactions	Turnover rate
Costs	Front load
Costs	Defer load
Costs	Expense ratio

Table 3.5. Typical raw input for self-organizing map formulation

Fund name	Ticker	Manager name	Tot ret YTD	Tot ret 3 Mo	Tot ret annlzd 3 Yr	Tot ret annlzd 5 Yr	Mstar risk 3 Yr	Std dev 3 Yr	Bear mkt decile rank	Mstar rating	Net assets $MM	Turnover ratio	Manager tenure	Expense ratio
AARP Growth & Income	AGIFX	Management Team	21.61	7.86	18.22	15.86	0	10.42	3	5	4622.5	25	10	0.69
American Capital Exchange	ACEHX	Boyd/Harrel	36.21	13.97	23.73	16.36	0	13.76	5	5	62.3	0	6	0.88
Davis NY Venture A	NYVTX	Davis/Davis	26.54	11.69	20.38	17.8	0	13.09	6	5	2697.1	19	27	0.87
Dodge & Cox Stock	DODGX	Management Team	22.27	8.87	19.7	17.6	0	12.12	7	5	2227.6	13	29	0.6
Elfun Trusts	ELFNX	Carlson, David B.	23.55	7.95	19.9	15.5	0	11.11	6	5	1526.1	15	8	0.13
Evergreen Growth & Income Y	EVVTX	Nicklin Jr., Edmund H.	23.82	6.96	18.74	16.88	0	10.81	5	5	433.8	17	10	1.27
Fidelity Adv Growth Opport T	FAGOX	Vanderheiden, George	17.73	8.82	17.23	17.75	0	10.68	8	5	15527.9	39	9	1.58
Fidelity Contrafund	FCNTX	Danoff, Will	21.94	8.49	18	18.25	0	12.04	6	5	23920.8	223	6	0.96
Fidelity Destiny I	FDESX	Vanderheiden, George	18.55	8.85	19.24	19.79	0	11.55	7	5	5088.7	42	16	0.65
Fidelity Destiny II	FDETX	Vanderheiden, George	17.86	8.62	18.74	19.64	0	11.33	8	5	2858.7	37	11	0.78
First American Stock A	FASKX	Management Team	29.1	9.96	21.04	17.08	0	10.75	4	5	27.7	52	9	1
First Eagle Fund of America	FEAFX	Cohen/Levy	29.34	10.54	19.78	21.48	0	13.67	4	5	174.9	81	9	1.9
Franklin CA Growth I	FKCGX	Management Team	30.43	9.03	30.92	22.58	0	14.69	NA	5	168.2	62	5	0.71
GMO Core III	GMO3	Management Team	17.61	6.86	19.92	16.27	0	12.1	4	5	3508.8	77	11	0.48
GMO Small Cap Value	GMOSM	Management Team	20.16	6.84	16.67	18.84	0	11.24	NA	5	368.6	135	5	0.48
GMO Value III	GMOV	Management Team	20.73	10.11	18.84	16.85	0	11.78	NA	5	21.5	65	6	0.61
Homestead Value	HOVLX	Teach, Stuart E.	17.94	5.26	17.38	16.5	0	10.64	NA	5	237.3	10	6	0.84
Kaufmann	KAUFX	Auriana/Utsch	20.91	-0.87	21.73	18.87	0	17.88	9	5	527.4	60	10	2.17
Kemper-Dreman High Return A	KDHAX	Dreman, David N.	28.79	9.5	23.18	19.63	0	13.74	9	5	351.6	18	8	1.25
Lazard Equity	LZEQX	Management Team	19.91	7.9	19.73	16.54	0	11.25	6	5	278.1	81	9	0.92
Longleaf Partners	LLPFX	Hawkins/Cates	21.02	6.73	18.9	19.88	0	10.09	4	5	2376.6	13	9	1.01
MainStay Inst Val Eq Inst	NIVEX	Kolefas/Laplaige	22.41	10.05	17.05	17.34	0	10.26	NA	5	815.6	51	5	0.93
Mairs & Power Growth	MPGFX	Mairs III, George A.	27.76	7.93	25.86	19.4	0	13.49	7	5	148.2	4	16	0.99
MAS Value Instl	MPVLX	Management Team	27.63	12	22.37	19.14	0	11.58	6	5	2279.3	53	8	0.6
Merrill Lynch Growth A	MAQRX	Johnes, Stephen	29.72	12.87	21.78	21.33	0	15.84	10	5	112.4	37	8	0.82
Merrill Lynch Growth B	MBQRX	Johnes, Stephen	28.38	12.59	20.54	20.09	0	15.71	10	5	3087.2	37	9	1.84
Morgan Stanley Inst Eqty Gr A	MSEQX	Feuerman/Johnson	31.14	8.26	25.23	17.03	0	12.1	NA	5	245.2	186	5	0.8
Muhlenkamp	MUHLX	Muhlenkamp, Ronald H.	30.08	11.96	17.09	17.04	0	11.35	8	5	39.9	23	8	1.35
Mutual Beacon Z	BEGRX	Price, Michael F. (et al)	21.19	7.54	17.23	19.48	0	8.26	2	5	4688.7	73	11	0.72
Mutual Qualified Z	MQIFX	Price, Michael F. (et al)	21.19	7.75	17.5	19.55	0	8.77	3	5	4095.6	76	16	0.72
Mutual Shares Z	MUTHX	Price, Michael F. (et al)	20.76	7.9	17.68	19.06	0	9.81	3	5	6185.5	79	21	0.69
Oakmark	OAKMX	Sanborn, Robert J.	16.21	7.6	17.3	25.68	0	10.58	NA	5	4196.1	24	5	1.18
PBHG Growth PBHG	PBHGX	Pilgrim, Gary L.	9.82	-6.91	20.04	26.64	0	23.36	10	5	5931.2	45	11	1.48
Pegasus Intrinsic Value I	WOIVX	Neumann/Gassen	23.97	8.49	15.29	15.07	0	7.65	NA	5	338.7	46	5	0.91
Pelican	PELFX	Mayo, Richard A.	20.69	9.51	17.33	16.72	0	10.52	4	5	186.1	40	7	1.1
Sequoia	SEQUX	Cunniff/Ruane	21.74	7.43	21.16	16.59	0	13.34	3	5	2677.8	15	26	1
Sound Shore	SSHFX	Burn III/Kane Jr.	33.68	12.01	20.31	18.76	0	10.69	5	5	116.8	53	11	1.15
Spectra	SPECX	Alger, David	19.48	1.54	22.3	20.41	0	21.77	9	5	17.1	NA	22	NA
Stepstone Val Momentum Instl	UIVMX	Earnest, Richard	25.64	9.45	19.57	16.08	0	11.41	NA	5	282.7	20	5	0.8
Strong Common Stock	STCSX	Weiss/Carlson	20.47	8.45	16.65	19.13	0	11.23	8	5	1243.6	92	5	1.2
Strong Schafer Value	SCHVX	Schafer, David K.	23.17	11.81	16.51	18.4	0	11.35	9	5	514.2	18	11	1.27

criteria we used were the management tenure equal or greater than three years, shareholders grading equal or greater than B+, Morningstar rating equal or greater than 4 (on a scale of 1 to 5), and/or an expense ratio equal or less than 1%.

3.4.3 Results from Applying SOM to Different Selections

The SOMs presented here demonstrate differences among mutual funds which Morningstar currently groups under the same category labels. Better information on the main differences between seemingly similar funds should help to improve fund selection as well composition of portfolios of mutual funds that better reflect the desiderata of investors.

3.4.3.1 World Stocks

A SOM of 50 mutual funds focused on world stocks, all with a Morningstar rating of four or higher, is shown in Figure 3.3. There are three groupings among these 50 funds:

- group I includes Colonial Global Equity B (COEBX), Dean Witter Global Div Growth (DWGDX), MFS World Equity B (MWEBX), MFS World Growth B (MWOBX) and Oppenheimer Quest Global Value B (QGLBX);

- group II consists of all remaining funds (the ticker symbols of which can be found in Morningstar's listing of international equity funds);

- group III includes New Perspective (ANWPX), Capital World Growth & Inc (CWGIX), Janus Worldwide (JAWWX), Oppenheimer Global A (OPPAX), Templeton World I (TEMWX), and Templeton Growth I (TEPLX).

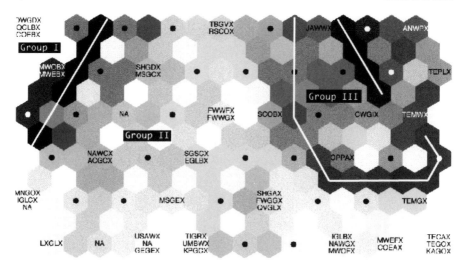

Figure 3.3. Self-organizing map of 50 mutual funds that are all focused on investments in world stocks. The data was derived from the Morningstar database (April 1997). The map shows three main groupings among funds that have Morningstar ratings of four or five stars. *(This figure can be seen in color in the Color Plate Section.)*

Some simple averages of the original input data for each of these groups of funds are shown in Table 3.6. From this table we note that

- group I funds have managers with few years of tenure (on average 2.8 years, as opposed to the managers of group III funds who have on average 7.2 years tenure);
- group I funds produced on average about 200 basis points per year less than the funds in group III (both the three- and the five-year annualized performances of funds in group I average about 200 basis points per year less than the funds in group III);
- the average net assets under management of funds in group I is about one tenth of the funds in group III;
- the turnover rate of funds in group I is much higher (82%) as compared with the turnover rate of funds in group III (which is about 52%);
- the expense ratio of group I funds is twice as high (2.3%) on average than the average expense ratio of group III funds (1%).

Table 3.6. Analysis of mutual funds focused on world stocks

Group	No. funds	Manager tenure	Tot ret 12 mo	Tot ret annlzd 3 yr	Tot ret annlzd 5 yr	Bear mkt decile rank	Net assets $MM	Turn-over ratio	Front load	Defer load	Expense ratio
1	5	2.8	19.0	13.8	14.1	4.0	658.2	80.8	0.0	4.6	2.3
2	39	3.3	20.1	14.1	14.6	6.7	272.4	70.7	2.2	0.1	1.7
3	6	7.2	22.4	16.1	16.0	5.8	6638.3	52.7	4.8	0.0	1.0

3.4.3.2 International Bonds

Two SOMs of funds focused on international bonds are shown in Figures 3.4 and
3.5. Figure 3.4 is a map of 36 international bond funds with Morningstar rating 4
or better. Figure 3.5 zooms in on a subset of 24 of these funds that have managers
with three or more years of tenure.

Both maps classify Standish International Fixed-Income (SDIFX) managed by
Kenneth Windheim in a separate category. Figure 3.4 also puts T. Rowe Price
International Bond (RPIBX), managed by Peter Askew, in group I. Figure 3.5 shows a
separate group consisting of Franklin Tax-Adv Intl Bond (FRIBX), managed by D.
William Kohli, Global Government Plus A (GGPAX), managed by Irwin Wells, and
Global Total Return A (GTRAX), also managed by Irwin Wells. The ticker symbols of
all the other funds can be found in Morningstar's listing of international bond funds.

What are the common features of these groups of international bond funds? First,
SDIFX and RPIBX are two of the largest funds in this category. They have currently
$920 and $980 million assets under management, respectively. In comparison, the
average assets under management by the funds in group II is only $158 million.
Second, the average three- and five-year annualized performance of SDIFX and
RPIBX is 100 to 200 basis points per year better that of the funds in group II. Third,
the average expense ratio of these two funds is 40 basis points lower than the average
of the others. Finally, the funds in group III in Figure 3.5 have front loads that are
about four times as high as those of all other funds on these maps.

3.4.3.3 Large and Midsize Stocks

The SOM shown in Figure 3.6 maps 122 mutual funds that invest in domestic or
US-based large and mid-cap (growth or value) stocks. All of the funds on this map

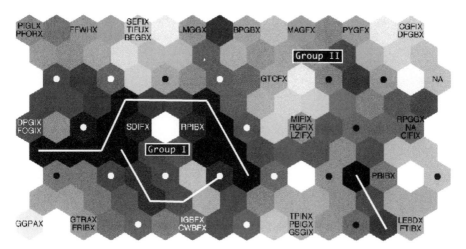

Figure 3.4. Self-organizing map of 36 mutual funds that are all focused on investments in international
bonds. The data was derived from the Morningstar database (April 1997). The map shows at least two
main groupings among funds that have Morningstar ratings of four or five stars. *(This figure can be
seen in color in the Color Plate Section.)*

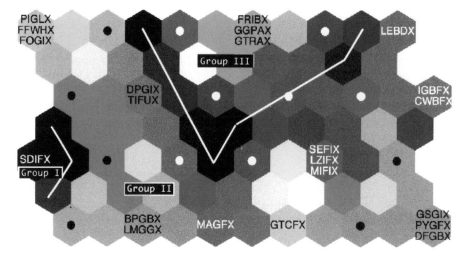

Figure 3.5. SOM of 24 mutual funds focused on international bonds, using map size 6 by 4.

have Morningstar ratings of four or higher and all managers of these funds have managed these funds for four years or more.

The SOM in Figure 3.6 shows three groups of funds. A summary analysis of the main differences between these three groups is provided in Table 3.7. Funds in group I have three- and five-year annualized returns that are 500 basis points higher than the average returns of the funds in group III. While the net assets under management are significant higher for group I than group III, we note that the turnover ratio and the volatility among these groups of funds is significantly different. The turnover ratio of the first group of funds is 81.1% versus 32.2% on average for the third group; the three-year average volatility of the first group of funds is 17.3% as opposed to 10.8% on average for the third group of funds. Finally, we also see that the deferred loads on the five mutual funds in the third group is much higher (4.8%) than the deferred load on all the other funds.

3.4.3.4 Small Cap Stocks

Some 35 mutual funds that invest in domestic small-cap stocks are mapped in Figure 3.7. This SOM shows two groups. Group II contains Franklin Small Cap Growth I (FRSGX), FPA Capital (FPPTX), Biltmore Special Values A (BTSVX). Heritage Small Cap Stock A (HRSCX), and Pioneer Capital Growth A (PCGRX); all other funds fall in group I. The ticker symbol of all the remaining funds on these maps can be found in Morningstar's listing of small cap funds.

The differences between these groups of small-cap mutual funds are shown in Table 3.7. The second group of funds are smaller in size, have higher returns, and substantially higher front loads than the first group. Figure 3.8 zooms in on 14 small cap mutual funds and maps these into funds into another three sub-groups.

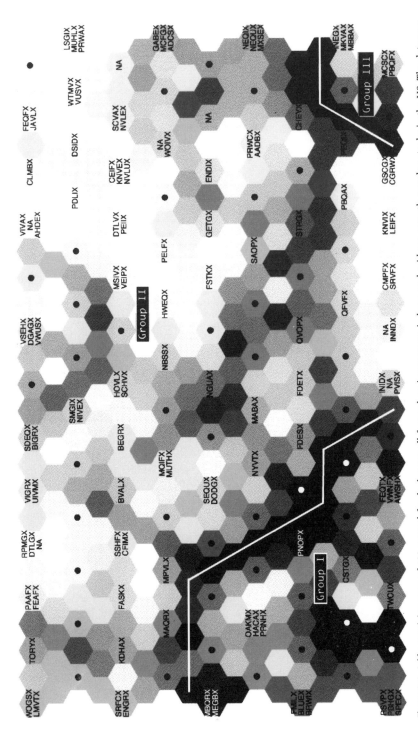

Figure 3.6. Self-organizing map of 122 mutual funds that are all focused on investments in large and mid-cap growth or value stocks in the US. The data was derived from the Morningstar database (April 1997). The map shows at least three main groupings among funds that have Morningstar ratings of four or five stars. *(This figure can be seen in color in the Color Plate Section.)*

Table 3.7. Analysis of mutual funds focused on domestic large and mid-cap vs. small-cap stocks

Group	No.	Manager tenure	Tot ret 12 mo	Tot ret annlzd 3 yr	Tot ret annlzd 5 yr	Bear mkt decile rank	Net assets $MM	Turn-over ratio	Front load	Defer load	Expense ratio
Large and mid-cap stocks											
1	17	9.2	20.4	22.3	22.6	8.7	7086.6	81.1	1.2	0.5	1.1
2	100	8.1	25.5	19.7	18.6	5.5	916.8	51.8	1.2	0.1	1.0
3	5	6.8	21.0	16.9	17.7	4.8	1919.6	32.2	0.0	4.8	1.7
Small cap stocks											
1	30	6.9	23.8	21.2	21.0	7.6	880.6	62.9	0.3	NA	1.2
2	5	5.6	28.9	23.1	24.2	10.0	528.9	62.8	5.2	NA	1.2

In comparison with the large and mid-cap mutual funds mapped in the previous section it is interesting to note that

- the average management tenure of small cap funds is smaller than that of large and mid-cap funds (although group III of the large and mid-cap funds is about the same as group I of the small cap funds);
- the three- and five-year annualized returns of the small cap funds is on average higher than the return of large and mid-cap funds;
- volatility of the small cap funds is higher than that of groups II and III of the large and mid-cap funds.

3.4.3.5 Emerging Markets

In all the examples provided so far we used SOM to create maps of mutual funds. From these maps we discovered groupings of funds. Simple statistical analysis demonstrated key differences among groups. In this final section we demonstrate how SOM maps can be turned into decision-support tools for selection of mutual funds as well as for the creation of benchmarks or indices for monitoring mutual funds.

As an illustration we will use a selection of mutual funds that focus on investments in emerging markets. For maps of individual emerging markets as well as individual companies in emerging markets we refer to Chapter 6.

To begin our search for patterns among funds that invest in emerging markets we start by creating a Sammon map. A Sammon map provides a rough visual inspection of the pairwise distances between data vectors. Sammon mapping tries to approximate local geometric relations of the data samples in a two-dimensional graphic plot. Sammon's mapping is particularly useful for preliminary analysis because it can be made to roughly visualize the class distributions and the degree of overlap.

Figure 3.9 shows the Sammon projection of some 82 mutual funds who invest in diversified emerging markets in Latin America, the Pacific, and European

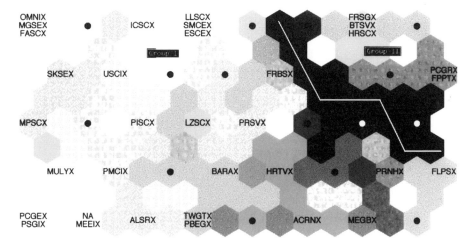

Figure 3.7. SOM of 35 mutual funds focused on small growth and value stocks, using map size 8 by 5.

countries. This Sammon mapping demonstrates that quite few funds are near to each other, while some are at far distances from the main clusters (Sammon, 1969).

Figure 3.10 shows the SOM map of the same 82 funds. All funds included in this selection have a shareholders grade equal or greater than B+ and have been managed by the same manager for at last the last three years. The SOM map presented in Figure 3.10 contains 12 by 8 neurons. The grayscale overlay of the map allows to identify at least five groups (possibly six). The main differences between these groups are tabulated in Table 3.8.

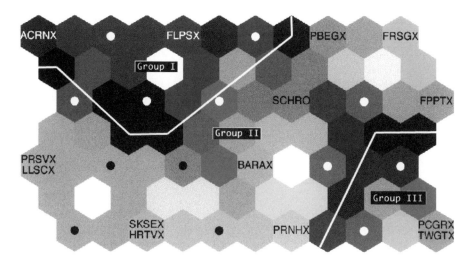

Figure 3.8. SOM of 17 mutual funds focused on small growth and value stocks, using map size 6 by 4.

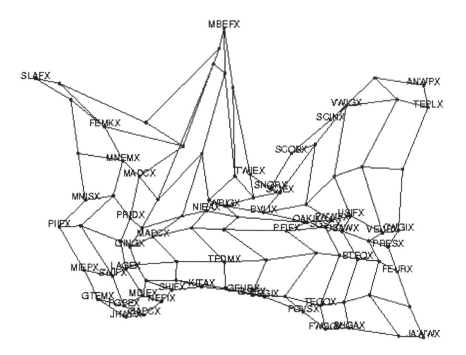

Figure 3.9. Sammon projection of 82 mutual funds focused on investments in emerging markets. A comparison of management tenure and turnover ratio shows a inverse correlation, meaning that funds managed by managers who have been less long at the helm also turn their portfolios around quicker (group I and III versus groups II, IV and V).

Funds in groups I and IV are larger in terms of assets under management ($1.4 to $7.9 billion); and they have on average higher returns (200 to 400 basis points higher) than the funds in the other groups. Interestingly enough, in this class of funds investing in emerging markets this is not correlated with management tenure, nor with turnover or expense ratios.

A more in-depth analysis of each input variable is provided by the *component plane representation*s. By component plane representation we can visualize the relative impact of the input data. Component plane representation can be thought of as a sliced version of the SOM. Each component plane has the relative distribution of one data vector component. In this representation, dark values represent relatively small values, while white values represent relatively large values. By comparing component planes we can see if two components correlate. Component planes can also be used for detecting correlation's among inputs that are different in different areas of the input space. If the image in the same part of two component planes is similar, then those inputs correlate strongly.

The component plane representation provides a clear visualization of correlation between the vector components. By picking the same neuron in each plane (in the same location), we could assemble the relative values of a vector of the network.

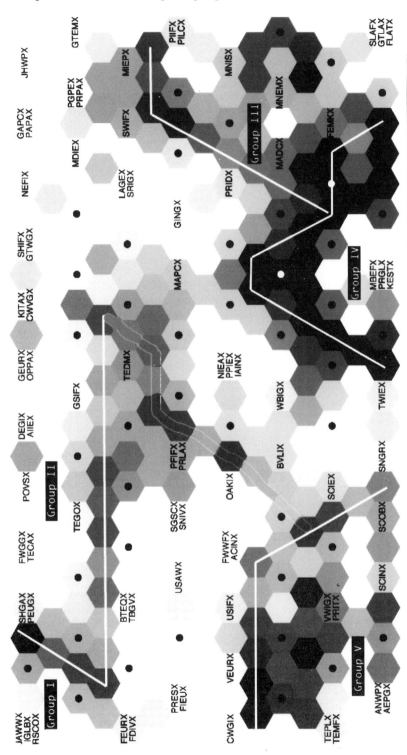

Figure 3.10. Self-organizing map of 82 mutual funds that are all focused on investments in emerging markets. The data was derived from the Morningstar database (April 1997). The map shows at least five main groupings among funds that have Morningstar ratings of four or five stars. (*This figure can be seen in color in the Color Plate Section.*)

Table 3.8. Analysis of mutual funds focused on emerging markets

Group	No.	Manager tenure	Tot ret 12 mo	Tot ret annlzd 3 yr	Tot ret annlzd 5 yr	Bear mkt decile rank	Net assets $MM	Turn-over ratio	Front load	Defer load	Expense ratio
1	5	4.4	28.9	17.4	15.9	5.0	1403.2	125.0	1.1	0.0	1.7
2	60	5.1	17.2	10.0	11.7	6.7	566.8	59.4	2.9	0.2	1.6
3	6	3.8	15.8	4.7	10.6	8.0	518.3	136.2	2.3	0.2	1.8
4	3	6.7	18.9	11.8	12.4	6.7	403.3	66.0	0.0	4.3	2.3
5	8	14.3	17.8	13.0	13.8	5.4	7930.7	24.3	2.9	0.0	1.0

A component plane representation of the emerging market mutual funds selected for the SOM in Figure 3.10 is shown in Figure 3.11. Each plane in this figure represent one input variable. For example, the first plane shows the SOM slice on management tenure, the last one shows the SOM slice on expense ratio.

From the component plane representation in Figure 3.11 it can be seen that

- management tenure does not overlap with the performance related slices: the highest values for three-year annualized return are in the upper left corner, while the highest management tenure values are in the lower left corner;
- the highest Morningstar ratings are all to the left of the map, while the lowest ratings can be found in the bottom right corner;
- the volatility of returns as measured by the standard deviation is highest for the funds in the bottom right corner and lowest for those to the left of the map;
- at the center as well as to the left of the map we find very low bear market decile rankings;
- the highest turnover ratios can be found in the middle right corner of the map, the lowest turnover ratios in the center and left part of the map;
- the front-, deferred and expense ratio SOM slices are each quite different: the highest ratios can respectively be found in the top left part for front loads, and the bottom right for deferred loads and expense ratios.

If the highest and lowest values of key variables are highlighted and brought together on a single map, we obtain what shall be called an *investment decision support table* or labeled *investment map* (similar to the solvency map discussed in Chapter 1). An example of such a mutual fund decision-support table is provided in Figure 3.12. This lookup table shows that mutual funds that invest in diversified emerging markets group in clusters along various parts or zones of this lookup table. For example, we note that funds with the highest Morningstar ratings cluster in the top left corner of the table and funds with the lowest Morningstar ratings cluster in the bottom left corner. Funds with large assets under management cluster in the bottom left corner, while funds with small assets under management cluster in the right hand top corner. Funds with the largest front loads can be found in the middle top of the map; those with smallest front loads at the bottom.

There are no sharp divisions between the zones in Figure 3.11, however, the decision support table provides a translation of a SOM into a lookup table that provides for easy

Figure 3.11. The component plane representation of the SOM on emerging market mutual funds.

selection and composition of mutual funds with different attributes. This labeled SOM is like a metro or subway map of a city which highlights the stations and the routes without providing the full detail of a city map. Like a subway map this labeled SOM can be used as a lookup table for deciding on the best route, path or destiny for achieving results with mutual fund investments.

3.5 Conclusions

In this chapter we applied SOM to various selections of mutual fund data from Morningstar data. Unsupervised neural network approaches were deployed to visualize and compress multi-dimensional data into two-dimensional

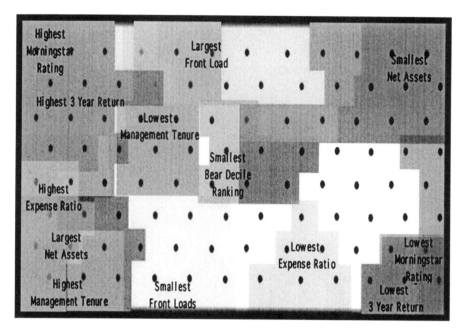

Figure 3.12. Labeled SOM of emerging market mutual funds including highest and lowest zones on selected indicators of performance, management tenure, and costs.

representations or maps. SOM provided insights into the similarities and differences among seemingly similar mutual funds. Sammon projections were applied to demonstrate the relative distances of the input vectors. A component plane representation of a SOM demonstrated how various input variables contribute to the overall mapping of funds. Finally, a generalized SOM map was constructed as a decision-support tool or lookup table for selecting of mutual funds, and the construction of benchmarks or indices for monitoring the performance of various subclasses of funds.

4 Maps for Analyzing Failures of Small and Medium-sized Enterprises

Kimmo Kiviluoto and Pentti Bergius

Abstract

There are several ways to approach the phenomenon of corporate failure. The classic approach is to look at failure as a *terminal disease* that manifests itself in the financial statements of a company two or three years before the actual failure. An alternative approach is to view the tendency of an enterprise to fail as a *chronic disease* that manifests itself only occasionally. The fundamental issue is the ability of an enterprise to adapt to changes in its environment. If management is not able to react quickly enough to external impulses, the enterprise will have a "failure disorder" that leads to an increased risk of failure. This failure disorder may be latent for several years, i.e. there is a period of several years during which the enterprise has a high probability of failure. Simple linear discriminant analysis which is commonly used to predict the failure of companies is not capable of finding the more delicate patterns in financial statements of companies. In this chapter Kimmo Kiviluoto and Pentti Bergius show how the Self-Organizing Map is an essential tool for analyzing the financial statements of companies in order to find much more subtle patterns that are indicators of failure disorder.

4.1 Corporate Failure – Causes and Symptoms

There are several ways to approach the phenomenon of corporate failure. The approach that has been most studied during the last three decades views the failure as a terminal disease of the company that manifests itself in the financial statements two or three years before the actual failure. As the failure gets nearer, the statements of the company steadily get worse. All failing companies are also considered to behave in a more or less similar manner. Therefore, failure prediction techniques developed utilizing this approach can be based on a simple linear discriminant analysis; for example, the widely used Z-analysis (introduced by Altman).

This study is based on an alternative failure hypothesis that has been developed by Pentti Bergius of Kera Ltd., a Finnish financing company. As Kera Ltd. specializes in financing and development of small and medium-sized enterprises (SMEs), the hypothesis is developed for SMEs, and it may not be directly applicable to large enterprises.

According to this hypothesis, the tendency of an enterprise to fail can be regarded as a chronic disease that manifests itself only occasionally. The fundamental issue is the ability of an enterprise to adapt to changes in its environment; if the management is not able to react quickly enough to external impulses, the enterprise will have a "failure disorder" that leads to an increased risk of failure. However, the disease may be latent for several years: even after the symptoms disappear, there is a period of several years during which the enterprise still has a high probability of a failure.

In practice, it has been found that an analysis of one year's financial statements is insufficient to give a reliable picture of the state of the company. This is the case when trying to detect the symptoms of the "failure disorder" as well – at least two consecutive years should be analyzed together. When this is done, the symptoms can be roughly characterized as falling into three different types:

- Type I: the profitability is low for two successive years, and during the second year, the solvency is also low; this symptom seems to be present in two-thirds of the previously analyzed failure cases.
- Type II: the profitability plunges very low for one year, also driving the solvency low for the same year; about one-quarter of the previously analyzed failures had this symptom.
- Type III: the profitability is low for one year and very low for the following year; this symptom is less frequent than the first two, being present in about one-tenth of the failures analyzed.

Together, these three types of symptoms seem to be present in some 85% of the analyzed failures – not necessarily immediately before the failure, but possibly with a few years between the appearance of the symptoms and the actual failure. During this latency period, the companies that had suffered from symptoms of type I or II still had a rather low solidity.

The main differences between the failure hypothesis outlined above and the classical approach are, according to the "Kera hypothesis":

- an increased failure risk can manifest itself in several different ways;
- the failure does not necessarily evolve linearly, with financial statements steadily weakening from year to year – the "failure disorder" may be latent for some years;
- consequently, it is possible that the symptoms may be detectable only several years before actual failure, so that the last known financial statements show no sign of increased risk

Because of these differences, the Self-Organizing Map (SOM) seems to be an essential tool for analyzing the financial statements. With the SOM, it is possible to find much more subtle patterns of behavior than with some of the more classical tools, such as linear discriminant analysis.

4.2 Self-Organizing Map as a Tool for Financial Statement Analysis

In this chapter we are mostly concerned about qualitative analysis of financial statements. For visualization purposes we extensively employ the SOM that is discussed in detail in Part 2. SOM component planes, which were explained in Chapter 3, display valuable information with gray-level coding. Likewise, Sammon's projection of the SOM, which was also introduced in Chapter 3, and U-matrix representations can turn out to be useful. The U-matrix representation captures the relative distances between the map units: the darker the color, the farther the unit weight vectors are from each other. Separate clusters in the data thus will have a darker band between them on the map; visually, this resembles a mountain range separating two plateaus on a geographic map [4.01].

4.2.1 Data Preprocessing

Financial indicators derived from financial statements were used for training the SOM presented in this chapter. Before we trained the SOM these financial indicators were pre-processed. We chose histogram equalization for the pre-processing, since it seemed to suit SOM slightly better than variance normalization. In fact, histogram equalization transforms the original arbitrary distribution of indicator values to a nearly uniform distribution.

The histogram equalization is performed for each indicator separately in the following manner. If we use a 100-bin histogram, and have N financial statements, from which we derive N values of indicator X, then the $N/100$ smallest values of X get replaced by value 1, the $N/100$ next smallest by value 2 and so on, and the $N/100$ largest values get replaced by value 100. If there are equal values in the original data, the bin sizes may be adjusted so that all equal values fit into the same bin, thus having equal value also after histogram equalization.

4.2.2 "Semi-supervised" SOM

A central question we address in this chapter is "what is the probability of failure given certain financial statements?" To find this dependence, we use SOM in a "semi-supervised" manner. This means that the shape of the map is determined by the data that can be found in the financial statements, and other attributes of interest are just carried along with the weight vectors so that they can be used later for visualization.

Specifically, assume that we are given a data vector that contains both the kind of information that can be derived from the financial statements, and also information that becomes available only later on, such as whether the company that gave the financial statement went bankrupt within a certain time after giving the statement. The map unit weight vectors then have correspondingly a "financial statement information" part and a "corporate state information" part.

Now the SOM winner unit is searched using only that part of the vector that contains the financial ratios derived from the financial statements, but the whole data vector is used for updating the map unit weights. For the weight updates, we used here a Gaussian neighborhood, as it seemed to form a smoother mapping than the other common choice, the bubble neighborhood, thus making the visual inspection of component planes easier.

As an alternative to the "semi-supervised SOM", we could have colored the attribute planes so that the color of each unit is determined by the attributes of the data vectors that are mapped to that unit. However, this way there would be random variations in the color because of the statistical noise. This is not the case with the "semi-supervised" approach described above, as the neighborhood function smooths out those random variations.

4.2.3 Trajectory Maps

The use of financial statements data from several consecutive years can be done in a straightforward manner by concatenating the yearly data vectors to form a single long data vector. A drawback appears with this approach: training SOM with such data vectors results in a map that is difficult to interpret. For instance, there is no natural coordinate system for the multi-year map, even though for a one-year map such a coordinate system can be identified (see Section 4.4).

We have tried to solve this problem by using two SOMs in a hierarchy. The first-level SOM is trained with yearly financial statements, so that for a given year a company can be positioned on the first-level SOM based on its financial statement for that year. The second-level SOM is then trained with the company's coordinates on the first-level SOM during two or three consecutive years, as illustrated in Figure 4.1. This way, each unit on the second-level SOM corresponds to a trajectory on the first-level SOM, capturing one typical pattern of change in a company's financial statements from year to year.

4.3 The Data

The data used in this study consists of small and medium-sized Finnish enterprises. The sample was selected from a collection of partial histories of Finnish SMEs on the basis of the line of business and size. Companies that typically deliver very large, multi-year projects were excluded, as their financial statements tend to show great variation in different phases of the project. Additionally, very small companies that could not be considered as "serious businesses", as well as those that were larger than medium-sized, were excluded from the sample. We also required that the history and state of the enterprise was well known: if there was no data available for a period longer than two years before the bankruptcy, or if the last known financial statements of a supposedly non-failed enterprise were very poor, the company was rejected from the sample. No data was rejected from the original population because of being "atypical", or looking like an outlier.

In the final sample, there were 11,072 financial statements. These were given by

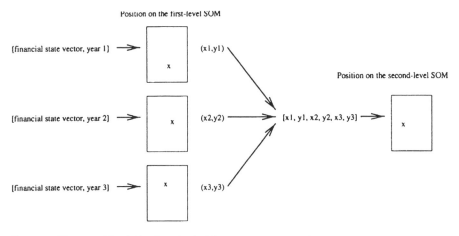

Figure 4.1. The second-level SOM is trained with vectors consisting of an enterprise's positions on the first-level SOM during two or three consecutive years.

2579 companies, of which 756 eventually failed, so there were 2606 financial statements that were given at most five years before failure.

The financial indicators chosen to train the first-level SOM were three commonly used ratios that measure the profitability and solvency of an enterprise.

4.4 Results

The first-level SOM is shown in Figure 4.2. Based on the financial indicators, the map positions itself so that its coordinates correspond to the solvency and the profitability: solvency increases from top to bottom, profitability from left to right. Notice how the area of highest proportion of corporate failures has a different location depending on the time to failure. Earlier the increased failure risk is mostly associated with low solidity, but later also with decreased profitability.

The Sammon's projection and U-matrix of the first-level SOM (Figure 4.3) reveal that the data is unimodal, i.e. not clustered. This is not a consequence of preprocessing the data using histogram equalization. This can be verified by examining the original marginal densities. There is one exceptional area on the map – the upper right-hand corner, as can be seen from the Sammon's projection and U-matrix. The capital structure makes the profitability measures – indicators I and II – behave differently in the upper right-hand corner than elsewhere.

In Figure 4.4, a few examples of company trajectories on the first-level SOM are shown. The trajectories generally tend to rotate clockwise: a decrease in profitability – a leftward movement – normally results in a decrease in solvency as well, which produces an upward movement. Exceptions to this rule indicate abnormalities, such as changes in the capital structure of the company.

The four leftmost trajectories are those followed by companies that eventually failed. The year the enterprise failed was mapped on each trajectory point and is

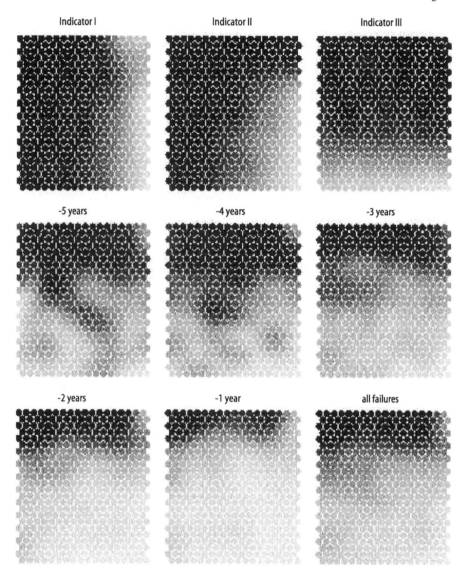

Figure 4.2. The first-level SOM. Light color corresponds to good relative values of financial indicators, or low proportion of financial statements given by failed companies.

plotted next to the trajectory; the area with high failure risk is marked with a (thresholded) darker shade.

The two-year and three-year trajectory maps are shown in Figures 4.5 and 4.6. In addition to the proportion of failures in different parts of the map, the three different failure types introduced in Section 4.1 are also indicated. As can be seen from Sammon's projection and U-matrix of the three-year trajectory map (Figure 4.7), the structure of this map is much more complicated than in the one-year

Sammon's projection   U-matrix

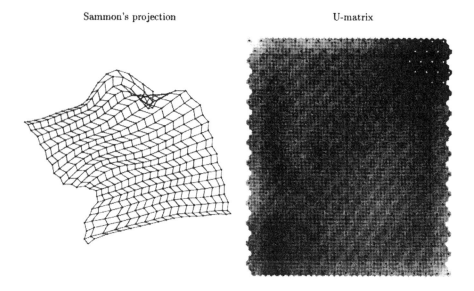

Figure 4.3. The Sammon's projection and U-matrix of the first-level SOM.

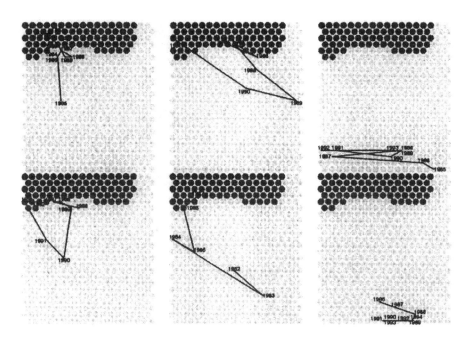

Figure 4.4. Trajectories of six companies on the first-level SOM.

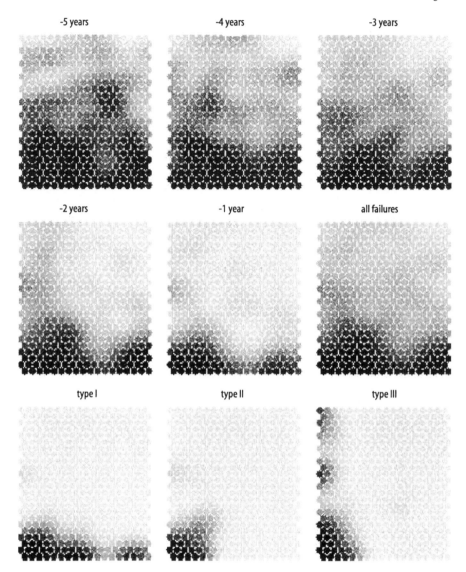

Figure 4.5. Two-year trajectory map.

case (Figure 4.3), but still no clearly separate clusters can be found. The Sammon's projection and U-matrix of the two-year trajectory map are very similar to those of the three-year trajectory map.

A very interesting way to look at the three-year trajectory map is displayed in Figure 4.8. Here, the first-level SOM trajectories that correspond to the selected units of the trajectory map are plotted on top of those units. The trajectories smoothly change throughout the map; the change would be even smoother if space would permit plotting all the units.

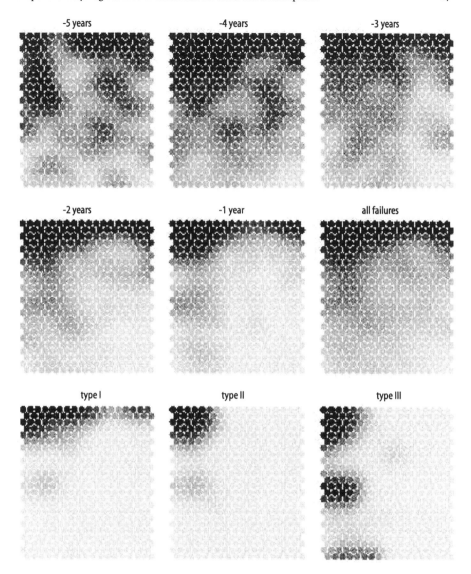

Figure 4.6. Three-year trajectory map.

A closer analysis of the second-level maps reveals that they capture information that escapes the first-level SOM. For instance, on the first-level SOM, the failing companies often jump out temporarily from the high-risk region. On the two-year trajectory map, however, there seem to be certain "absorbing states" or areas that failing companies generally do not leave. This behavior is summarized in Table 4.1: the absorbing states seem to be states 1 and 2, and the other states can be regarded as routes on which the companies enter the absorbing states. A similar statement also holds true for a group of very well performing companies. Because of

Sammon's projection   U-matrix

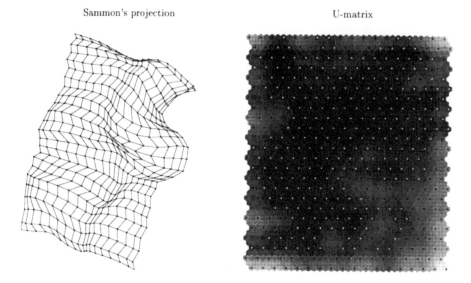

Figure 4.7. The Sammon's projection and U-matrix of the three-year trajectory map.

this property, the trajectory maps seem to be a promising tool for rating of enterprises.

Finally, we show the number of failed and non-failed firms in each of the three maps (Figure 4.9). If these maps were used as a basis for binary classification, the accuracy would be practically the same in all cases. The training set classification results that can be used to get a rough idea of the true accuracy on an independent test set are shown in Table 4.2 [4.02]. However, this kind of failed/non-failed classification is not very useful in practice: it simplifies matters too much, as there is no difference made between an excellent company and another that is only very little better than the best of the companies that are classified as failing. There is also the problem of choosing the time when a failing company changes from a healthy one to a failing one. Here, the time has always been chosen as five years before the actual failure, regardless of the shape of the company at that time, even though a better choice would probably be to change the state when the failure symptoms first appear.

Table 4.1. Transitions between certain areas of the two-year trajectory map that are associated with increased risk of failure. Transitions have been calculated from the point when a failing enterprise first enters one these areas.

To	1	2	3	4	5	Other	Terminates
1	168	11	1	2	10	13	51
2	13	20			2	1	9
3	4	4			2	1	16
4	14	4			5	2	15
5	16			2	9	7	31

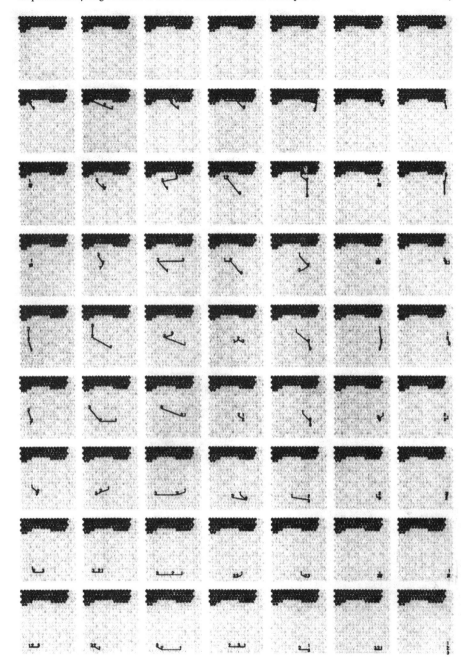

Figure 4.8. The three-year trajectory map "opened": on top of the map units are plotted the corresponding first-level SOM trajectories (about every third unit shown).

One-year map

Two-year trajectory map

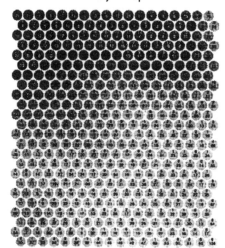
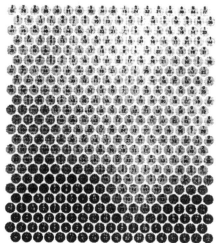

Three-year trajectory map

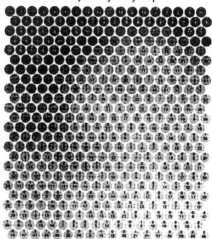

Figure 4.9. Number of failed (upper figure) and non-failed companies (lower figure) on the first-level SOM and two second-level SOMs. Here a company has been considered as failed if there were at most five years of financial statements that were used to determine its position on the map to the actual failure.

Table 4.2. Classification results for type I error target values 23.0% and 13.0%. Here type I error means classifying a failing company erroneously as a healthy one, type II error means classifying a healthy company erroneously as failing

Map	Total error	Type I error	Type II error
First-level SOM	20.2%	22.5%	19.6%
Two-year trajectory SOM	19.9%	23.3%	19.1%
Three-year trajectory SOM	19.8%	23.7%	19.0%
First-level SOM	28.0%	13.5%	31.6%
Two-year trajectory SOM	28.2%	13.0%	31.8%
Three-year trajectory SOM	27.0%	13.5%	30.0%

4.5 Summary

The Self-Organizing Map was shown to be a valuable tool for analyzing financial statements. In this study, the emphasis was on corporate failures, but the methods prove to be directly applicable to non-failed companies as well. The results seem to support the hypothesis developed in the first section, according to which there are several different types of corporate failures, and some of these cannot be detected on the basis of the last two or three years' financial statements only. It was shown that with SOM it is possible to recognize different patterns of corporate behavior and find attributes associated with those patterns. This makes the SOM look like a promising tool for a more general analysis of financial statements. In particular, applying SOM for corporate self-benchmarking and corporate rating seems feasible.

5 Self-Organizing Atlas of Russian Banks

S.A. Shumsky and A.V. Yarovoy

Abstract

The banking system in Russia is in deep crisis. Since 1994 the number of working banks has constantly decreased. Financial analysts predict even more drastic decreases in the future. In this chapter Serge Shumsky and A.V. Yarovoy present an analysis of newly available data on the emerging Russian banking system. They develop a methodology that involves the use of principal component analysis and unsupervised artificial neural networks in order to extract useful information from this newly published data. They discuss the qualitative meaning of different approaches and pay special attention to estimating the limitations of their results. This chapter demonstrates the value of unsupervised artificial neural networks and self-organizing maps in particular, as a tool for financial analysis of banking institutions.

5.1 Introduction

The banking system in Russia is in a deep crisis. Since 1994 the number of working banks has constantly decreased; financial analysts predict more drastic decreases in the near future [5.01]. To understand the reasons for the current crisis an in-depth analysis is desirable. Until recently, complete and reliable data about the financial state of Russian banks was not publicly available. In December 1996 the Central Bank of Russia published for the first time a collection of annual balance sheets and income statements of all Russian banks for fiscal years 1994 and 1995 [5.02]. In the near future this will be supplemented by similar information for 1996. This published accounting data implicitly characterizes the state of contemporary Russian banks (Altman, 1968).

Modern financial analysis has a wide arsenal of analytical tools [5.02]. However, the peculiarity of Russian accounting and Russia's unique economic situation makes the experience of Western economies in analyzing banking problems of limited applicability. Nobody has so far been able to extract a set of significant financial indicators that characterize the current financial status of banks in the rapidly evolving Russian economy. This deficit benefits, however, the artificial neural network (ANN) approach (Trippi and Turban, 1993; Deboeck, 1994; Refennes, 1995).

ANN is a technique that permits to extract from raw data explicit rules and models based on learning or successive approximations. Usually, supervised neural networks are applied to extract financial rules from data. The same technique can, however, not be applied to Russian banks, because of the extremely short history of available statistical data. This scarcity of historical data makes the unsupervised neural techniques the more desirable analytical tool for the analysis of the Russian banking data. Unsupervised ANN do not need an objective teacher, i.e. business failure records, in order to extract useful information from the available accounting data on Russian banks.

The objective of this chapter is two-fold. First, we will present an in-depth analysis of newly available data on the emerging Russian banking system. Second, we develop a methodology for the application of unsupervised ANN to the banking sector. Our goal is to reduce the data with minimum information loss. We first deploy principal component analysis for linear data preprocessing. Next, we deploy a nonlinear extension, the self-organizing map (SOM) algorithm for optimal data reduction and visualization of Russian banking data. We extend this data analysis by using various colorings of the map and show a *self-organizing atlas of Russian banks*.

5.2 Overview of the Russian Banking System

The number of Russian commercial banks increased until 1994. Since then the number of banks has constantly decreased, as illustrated in Figure 5.1. According to the Analytical Center of Financial Information the decline will continue more rapidly in the near future and only about 600 Russian banks are expected to survive by 1999. In 1999 the lower limit of each bank's own capital will be increased by the Russian Central Bank to 5 million ECU.

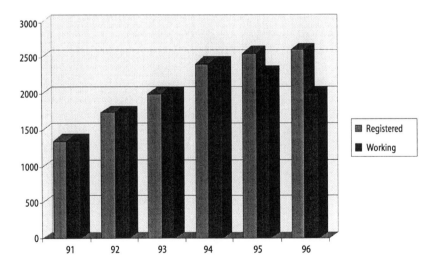

Figure 5.1. Number of registered and actually working Russian banks.

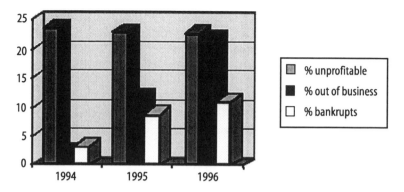

Figure 5.2. Profitability of Russian banks.

Figure 5.2 shows that almost a quarter of all registered Russian banks fail to be profitable in the present economic circumstances. As a result, more and more of them either close business voluntarily or are closed by the creditors. At the beginning of 1997 about 22% of all registered banks were already out of business, and the number of bankruptcies is still increasing. In 1996 12% of all registered banks went bankrupt. They accumulated about 20% of all bank debts.

Figure 5.3 shows that the structure of banks' assets has changed dramatically since 1993. Russian banks obviously favor government securities, while at the same time reducing their reserves and investments in the real economy. This is easily understandable given the typical 80–100% earning power of government securities in recent years. The Russian government is now constantly reducing the yield of these securities. Thus, in the near future, the structure of banks' assets may change. However, investments in the economy also require more concentration of banking capital. All these factors cause the instability of contemporary Russian banking.

The data that we used for the analysis includes 32 indicators on each bank – items in balance sheets and income statements. After filtering suspicious outliers, we retained 1794 input vectors for the year 1994 and 1780 vectors for the year 1995.

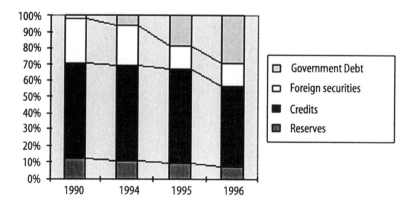

Figure 5.3. Structure of assets.

All data items were normalized based on the total assets of each bank because we are interested in relative, rather than absolute figures. Taking into account that total assets balance, i.e. total liabilities and equity, one then comes to 30-dimensional, normalized input data.

5.3 Problem Formulation

The ultimate goal of our analysis of Russian banks is to measure their relative health. One-dimensional ordering of banks, which is a widely used technique, has two general problems. One-parameter descriptions, though intuitively appealing, may be too coarse for optimal decision making. Nobody knows the algorithm for an objective rating of banks in various economic situations.

To overcome the first drawback of one-dimensional ordinal ratings, a two-dimensional ordering of banks is considered. Such a two-dimensional rating is capable of providing much more useful information than is readily understandable via simple visualization techniques [5.03]. However, two-dimensional ordering may also need objective coordinates similar to one-dimensional ratings. A common approach – using two "leading" financial indicators (e.g. Capital over Assets and Profit over Assets) obviously lacks the desired objectiveness. Ideally, all financial items should somehow contribute in representing the bank's status.

From a mathematical standpoint, visualization of complex multi-dimensional data can reduce the data dimension, i.e. perform data compression with a minimal loss of information. To put it in another way, our goal is to construct a smaller number of new variables that are functions of the initial data items, but to maximize their information content. If these new variables are constrained to be a linear combinations of the old ones, the solution is given by principal component analysis (PCA). On the other hand, a nonlinear PCA seeks to extend the results to arbitrary dependencies. Kohonen's self-organizing maps (SOM) represent the latter approach.

5.4 Linear Analysis using PCA

Geometrically speaking, PCA aims to find a hyperplane that best fits the given data. Amari (1992) and Linsker (1988, 1991) pointed out that PCA is equivalent to maximizing the information content in the outputs of a network of linear units. The advantage of neural PCA over the conventional statistical technique is that there is no need to compute the full correlation matrix. The network computes its first eigenvectors adaptively and directly from input data. This significantly saves computational effort in the case of extensive data compression.

PCA seems to be a reasonable first step in data analysis due to its simplicity and the clarity of the results. It finds a linear combination of the first eigenvectors, rather than giving the first eigenvectors themselves. Linear PCA presents a distinct measure of data compression quality. The error, minimized by PCA, is the sum of distances between the data points and their projections to the hyperplane. Normalized by total data dispersion, this quantity indicates the relative discrepancy caused by linear compression. The complementary quantity – data dispersion in the hyperplane over the total dispersion in raw data – is maximized by PCA.

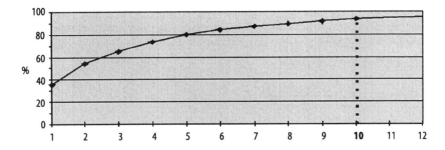

Figure 5.4. Percentage of total dispersion in the first principal-components hyperplanes.

The resulting plot for Russian banks in 1994 is presented in Figure 5.4. The ten first principal components contain 94% of total dispersion [5.04]. Using a controllable and robust linear PCA technique as the first step of data compression, we simplify the more comprehensive nonlinear compression stage.

Another result of PCA is the estimate of relative significance of the inputs. Figure 5.5 presents the relative precision of reconstruction of the data inputs by using the first ten principal components. The higher the bar the larger the significance of corresponding input in principle components, i.e. the more representative this input is in analyzing Russian banks. This technique allows for objective extraction of most valuable financial ratios without human expertise.

Since the error for two principal components is $E_{\mathrm{L}} = 0.47$, PCA alone is not able to compress data to two dimensions, which is necessary for a proper visualization. Hence, one is forced to use a more advanced nonlinear compression technique.

5.5 Nonlinear Analysis or the Nonlinear PCA Extension

Nonlinear PCA maximizes the information content in the output of nonlinear networks. A common way is to learn the outputs of a multilayer perceptron with a bottleneck hidden layer to reproduce its inputs (Cottrell et al., 1993). Error minimization in this case forces the bottleneck neurons to pass through as much information as possible. Thus, these neurons represent the optimal compression of input data.

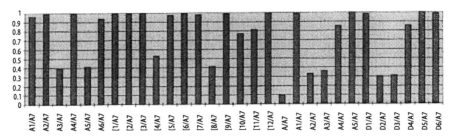

Figure 5.5. Quality of reconstruction of accounting items from the first 10 principal components.

This general approach, however, has a number of drawbacks. First, there are no theoretical grounds for the optimal number and sizes of hidden layers. Second, the learning of such multilayer perceptrons is very time consuming. And last but not least, the learning process may easily be stuck in numerous local minima of the error function. Thus, the optimality of the obtained compression is always dubious.

Another branch of data compression methods is represented by a family of winner-take-all (WTA) networks. Here, instead of minimizing the number of nodes in a bottleneck, one uses a larger number N of output neurons. But due to their intrinsic interaction only one of these neurons may fire in response to a network's input. Such sparse coding limits the amount of output information to $\log N$. This is chosen to be much smaller than the information content $d \log \varepsilon^{-1}$ of d-component input data measured with precision ε. So, such WTA networks actually compress information.

The *optimal* compression implies that the distribution of firing output neurons is uniform over the data set. This maximizes the information content of the output signal.

Contrary to the bottleneck networks described earlier, WTA networks may be implemented by a single layer of neurons. This greatly facilitates the learning procedure, though usually at the cost of heavier memory requirements.

SOM represents a topology-preserving variant of WTA networks. SOM produces a structured D-dimensional grid of output neurons, where close outputs correspond to close inputs. This property makes SOM an ideal tool for visualization of multi-dimensional data. Qualitatively, SOM maps d-dimensional input data to a D-dimensional output grid, preserving as much information as possible.

5.6 SOM of Russian Banks

For the sake of data visualization we used a two-dimensional grid of neurons fed by a 10-dimensional output of Oja's preprocessing PCA-network (Figure 5.6).

The layers of this hybrid two-layer neural network were learned one after another using the whole set of input vectors for a particular year. After the learning, each bank represented by a 30-dimensional input vector was processed by this

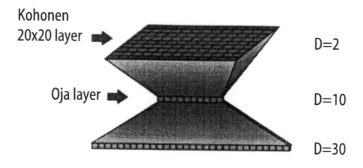

Figure 5.6. Two-layer neural network for visualization of accounting data.

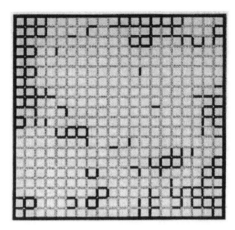

Figure 5.7. U-matrix display for the year 1994.

network and associated with a winner output neuron, i.e. each bank received its position on the resulting *SOM map*. The closer the positions of two banks on such a map the more similar are their financial profiles.

We used a 20×20 rectangular grid, so that the mean number of banks belonging to each cell was about four. In this case due to occasional fluctuations there were a couple of empty cells, though the SOM learning rule tends to uniform the output distribution. Increasing the number of output neurons is undesirable, resulting in a larger number of irrelevant neurons. On the other hand, choosing a smaller number of units decreases the resolution of the map.

The SOM algorithm combines the goals of both the projection and clustering algorithms. The map can be used to visualize the clusters in the data, using a U-matrix display (Ultsch, 1993a). In Figure 5.7 the more intensive the borders between the map cells the more the distance between corresponding grid nodes in the input space.

No distinct large clusters are noticeable in Figure 5.7. The internal nodes of the map are closer to each other than the nodes at the outer regions of the map. Thus, the data under consideration resembles the distribution of stars in a galaxy: there is a dense core surrounded by rarefied outskirts. The plot for the year 1995 is quite similar.

PCA provided a distinct measure of the quality of data representation by a given number of principal components. A similar measure may be introduced for nonlinear case as well; consider the ratio of the sum of squared distances between the inputs of the SOM layer and related neurons to the total data dispersion.

This measure is a straightforward generalization for nonlinear compression of the error for the linear case. The result for the data set of the year 1994 is $E_{NL} = 7.8\%$ (8.5% for the year 1995). Thus, the precision of nonlinear and linear ($E_L = 6\%$ for 10 principal components) layers are comparable, i.e. our network is properly balanced. The overall error of the mapping is less than 14%. This quantity is to be compared with a 47% linear error in the case of using two first-principle components for representing data.

5.7 A SOM Atlas of Russian Banks in 1994

The practical usefulness of the SOM map is due to its ability to visualize various items. In general, each balance item generates a certain coloring of the map. Figure 5.8 illustrates the origin of such coloring.

The cells of the map corresponding to greater values of a given item are colored more intensively. A set of all possible colorings, induced by all input items constitute what we call a *SOM atlas*. A subset of such a SOM atlas of Russian banks for the year 1994 is presented in Figure 5.9.

We limited ourselves to a single year since the data distribution of non-stationary Russian banking was quite different, even for consecutive years. Note the relative regularity of the coloring induced by the bank's size (see the Assets map). Recall that the network actually does not "see" this item, which was used for normalization. Nevertheless, the banks close in size appear to behave in a similar manner, e.g. the largest banks are concentrated in a very compact area. (Due to the large dynamic range of the bank size this map presents the logarithm of total assets.)

One can also recognize two separate types of small banks in the upper left and lower right corners of the map. The former are rather successful profitable banks with minimum liabilities, while the latter are not profitable at all (see the Profit coloring). In fact they do not grow – according to the Starting Capital coloring.

This type of bank, brought to life by the modern Russian economy, may be considered as "quasi-commercial". Their goal is not to maximize profit, but to serve as accounting branches of some larger enterprises.

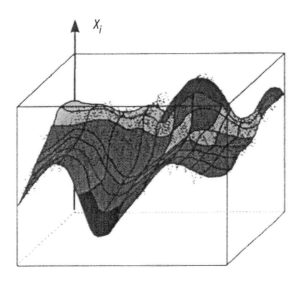

Figure 5.8. Coloring of the map, induced by the i-th item.

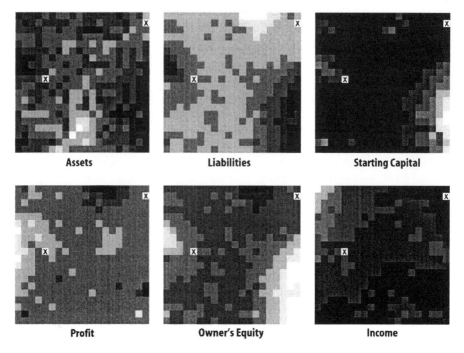

Figure 5.9. Part of the 1994 SOM atlas of Russian Banks.

Figure 5.10 presents the distribution of bankruptcies in 1994. It represents a subset of all bankrupts who nevertheless presented their financial reports. Note that both the above-mentioned groups – profitable and quasi-commercial banks – have not a single case of bankruptcy. Most bankruptcy cases are concentrated in the upper right corner of the map. This corner, according to Figure 5.10, is characterized by minimal profit and maximal liabilities, a dangerous combination indeed.

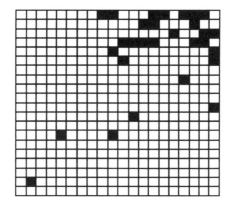

Figure 5.10. Distribution of bankruptcies in 1994.

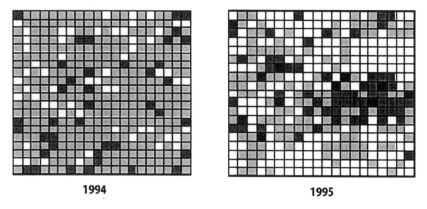

1994   **1995**

Figure 5.11. Density distributions of data on the map trained by 1994 data.

Thus, the correlations between various colorings in the Kohonen atlas may shed some light on the reason for bankruptcies and the existence of different types of reliable banks. Moreover, the relative area of the map under these groups gives a vivid and correct estimate of their relative role in modern Russian banking.

5.8 Evolution of Russian Banking from 1994 to 1995

As already mentioned, the distribution of data for the years 1994 and 1995 appears to be quite different. This can be clearly seen when the density distributions of both sets are represented in the same map (Figure 5.11).

The distribution of the 1994 data is relatively uniform (a property of the SOM algorithm). On the contrary, the data for 1995 is quite irregular on this map. The reason is an intensive migration of banks over the map. The mean shift of a bank's position is about seven cells (out of a maximum of 20). Such a turbulent dynamics prevents one from using the same map for visualization of data for different years.

A possible way out is to construct a new map for every year which somehow inherits the properties of its predecessor. This may be done by using the map of the previous year as an initial condition for the new one.

Figure 5.12 presents the resulting subset of the SOM atlas for the year 1995 similar to the one for the year 1994. Comparison of complementary figures allows one to analyze the tendencies in Russian banking. The total assets maps evidence the emerging diversification of the financial behavior of major banks. The gap between profitable and non-profitable banks became more pronounced, while non-commercial banks became more profitable.

5.9 Conclusions

The aim of this chapter was two-fold: to present the newly available data about Russian banking and to demonstrate a technique for data visualization and in-depth financial analysis.

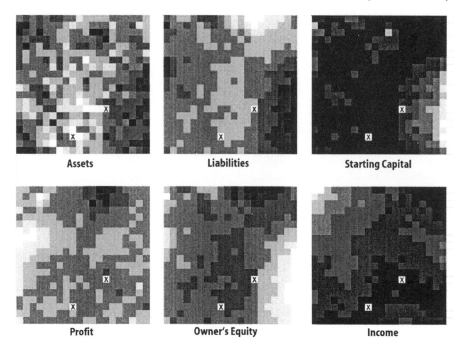

Figure 5.12. Part of the 1995 SOM atlas.

To make this novel technique more appealing, we concentrated on its optimality properties, tried to illustrate it in qualitative geometrical terms, and introduced recipes for quality control.

Two-dimensional representation of financial data seems to be more powerful than common one-dimensional ratings of banks. It provides much more information for financial analysts, and facilitates the inference of existing irregularities in large sets of complex financial data.

Color Plates

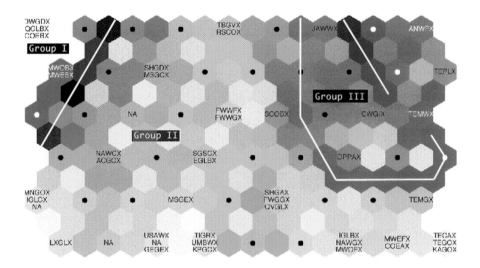

Figure 3.3. Self-organizing map of 50 mutual funds that are all focused on investments in world stocks. The data was derived from the Morningstar database (April 1997). The map shows three main groupings among funds that have Morningstar ratings of four or five stars.

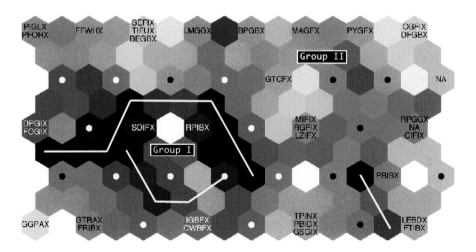

Figure 3.4. Self-organizing map of 36 mutual funds that are all focused on investments in international bonds. The data was derived from the Morningstar database (April 1997). The map shows at least two main groupings among funds that have Morningstar ratings of four or five stars.

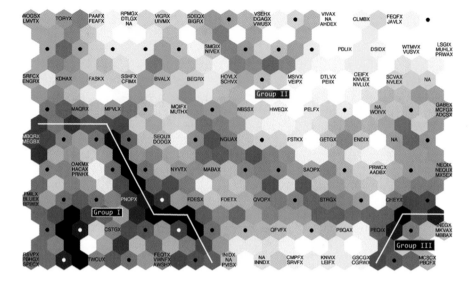

Figure 3.6. Self-organizing map of 122 mutual funds that are all focused on investments in large and mid-cap growth or value stocks in the US. The data was derived from the Morningstar database (April 1997). The map shows at least three main groupings among funds that have Morningstar ratings of four or five stars.

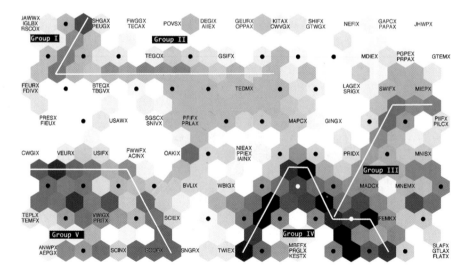

Figure 3.10. Self-organizing map of 82 mutual funds that are all focused on investments in emerging markets. The data was derived from the Morningstar database (April 1997). The map shows at least five main groupings among funds that have Morningstar ratings of four or five stars.

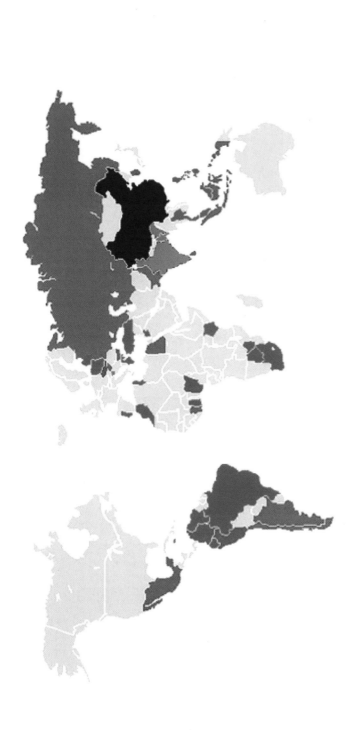

Figure 6.10. Self-organizing map of emerging and frontier equity markets grouped into five clusters, including:

- group 1: China, with a very high P/E of 30, high P/B ratio, low dividend payments, and moderate-sized companies ($189 million capitalization per company);
- group 2: Czech Republic, Slovakia and India, with small P/Es and low P/B ratios, and a large number of very small companies ($47 million capitalization per company) on average;
- group 3: eight countries, including Indonesia and Malaysia, with average P/E ratios of 21 and high P/B ratios, and low dividend payments on average;
- group 4: 11 countries, mostly in Latin America but also including Russia, moderate-sized companies ($160 million capitalization per company), low P/E (12), very low P/Bs and moderate dividend payments on average;
- group 5: 21 countries, mostly in Africa, with P/E ratios of 14, high P/B ratios, and relatively high dividend payments.

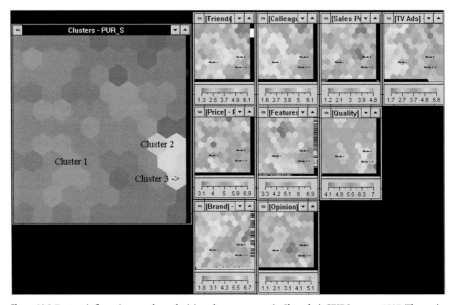

Figure 10.1. Factors influencing purchase decisions by consumers in Beijing, CEIBS survey 1997. The main map shows six clusters; the component planes on the right show the influences of various variables, e.g., advice of friends, TV ads, advice of colleagues, sales person, the product features, price, quality, brand name, the opinion of others. Blue areas in these planes indicate lower values; the red areas indicate the higher values.

Figure 10.2 Factors influencing purchase decisions by consumers in Shanghai, CEIBS survey 1997. The main map shows three clusters; the component planes on the right show the influences of various variables, e.g., advice of friends, colleagues, sales person, TV ads, price, features, quality, brand name and the opinion of others. Blue areas in these planes indicate lower values; the red areas indicate the higher values.

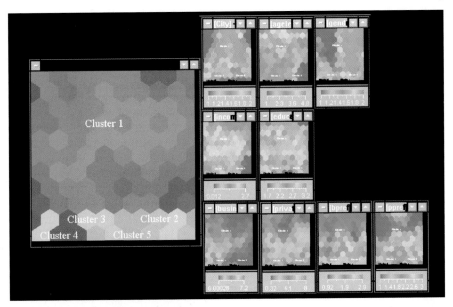

Figure 10.5. Self-organizing map of patterns in dining of Chinese consumers in Beijing and Shanghai, CEIBS survey 1997. The right side of the figure shows five main clusters among over 900 responses on this stratified survey. The component planes show the influences of locations (city: Beijing citizens, blue; Shanghai citizens, red); age (blue is younger, red is senior); gender (blue shows females, red shows males); income (blue shows no income, red shows more affluent); and education (blue shows little, red shows college education) on the number of business and private dinners (where blue areas show lower and red areas show higher frequency of dinners per month), and the cuisine preferences (where blue shows preferences for Chinese restaurants and red shows preference for Western restaurants).

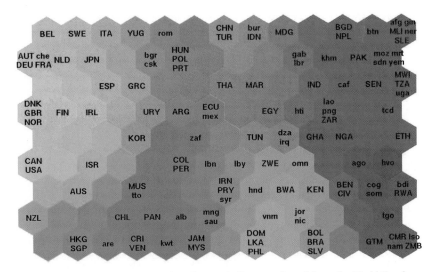

Figure 14.2. A SOM of world poverty, based on 39 indicators selected from the World Development Indicators published by the World Bank. In this map different colored cluster areas represent different poverty types. The color changes gradually among the clusters; similarity of poverty types is reflected in the similarity of colors.

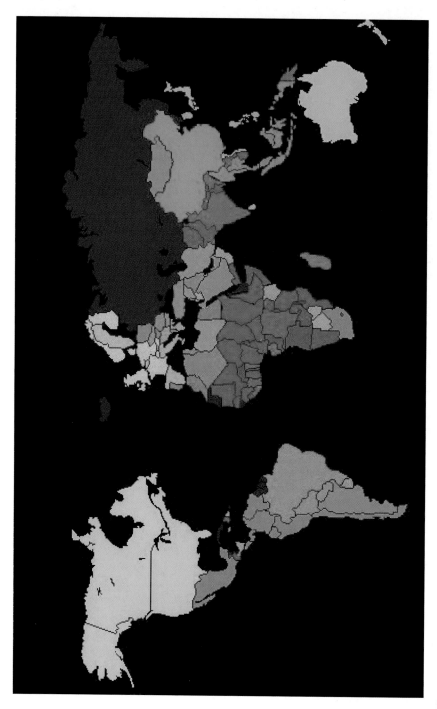

Figure 14.3. A world map based on poverty types obtained *via* a self-organizing map applied to 39 indicators selected from the World Development Indicators.

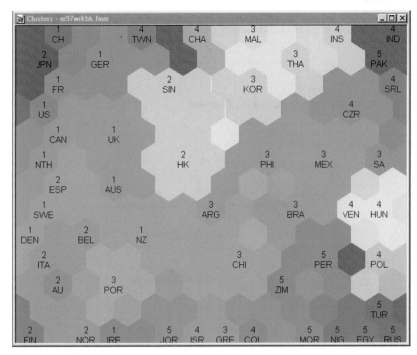

Figure 15.9. SOM map with labels on country credit risks derived from data published in the Wall Street Journal of June 26, 1997.

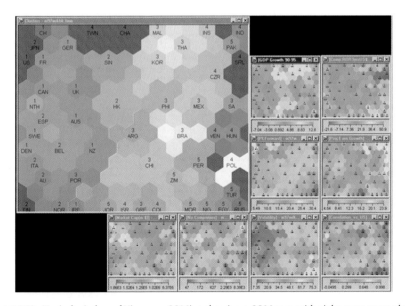

Figure 15.10. Typical window of Viscovery SOMine showing a SOM map with eight component planes allowing to derive non-linear dependencies among variables contributing to the clustering of countries based on country risk measures. Data used for this map was published in the Wall Street Journal of June 26, 1997.

for a free demo visit http://www.eudaptics.co.at

6 Investment Maps of Emerging Markets

Guido Deboeck

Abstract

The focus in this chapter is better understanding of the trends and patterns among today's emerging markets. About 80 countries currently have functioning securities exchanges. Based on the emerging stock market data published by the International Finance Corporation (IFC), this chapter analyzes patterns across emerging markets and patterns in the evolution of emerging stock markets over time. Fundamental and technical indicators are used to create self-organizing maps of the new markets in Latin America, Asia, Africa and Eastern Europe. In addition maps are provided that compare individual companies around the globe; maps of banking, telecommunication, and construction companies are included. The strategic importance of using self-organizing maps for investing in emerging markets is explored and the results are compared with those obtained through classical asset allocation strategies.

6.1 Background

According to IFC 80 countries in the world have functioning securities exchanges. Over the past ten years IFC has published an *Emerging Stock Market Factbook* which includes economic and stock-market data on the leading stock markets of the developing world. The IFC *Emerging Stock Market Factbook* is derived from the Emerging Market DataBase. Time-series data over ten years exist for about 30 emerging markets (IFC, 1996a, b).

Ten years ago the term "emerging markets" did not exist in the investment literature [6.01]. Today, with a vast array of information and services available on emerging markets, they often remain poorly understood. For example, a recent newspaper article pointed out that in the United States of America there are still investors that consider Hong Kong to be an "emerging market". Looking back into history, investors tend to forget that at one point in time the United States of America was an emerging market.

The emerging market economies on which IFC has substantial time-series data contain 4.8 billion people or 84% of the world's population. Of these 4.8 billion people, 2.7 billion live in the rapid growing economies of Latin America, China, India and Indonesia. In comparison, the developing countries contain 900 million people or 16% of the world's population. Various individual markets fall into different stages of development. The first generation of emerging markets – Japan, Hong Kong and Singapore – have already become developed industrial societies. Malaysia, Thailand, Indonesia and South Korea are considered the East Asia "Tigers" or new rapidly growing economies. In recent years Pakistan, India and Sri Lanka have joined in this trend and are called the "Young Tigers". Since the mid 1980s Latin American countries such as Argentina, Brazil, Chile, Mexico, Venezuela, Peru and Columbia have offered attractive investment opportunities. These emerging markets constitute a nonuniform group of countries. In southern Europe, Portugal, Greece and Turkey have opened negotiations with the European Union. In eastern Europe, Poland and Hungary have become emerging markets. The emerging markets constitute a rather highly differentiated group of countries, more so than the developed countries of the world.

At the end of 1996 the total market capitalization of all emerging markets amounted to 1.9 trillion US dollars. This figure represents approximately 12% of the 15.1 trillion US dollars total market capitalization of all world stock markets. In the period 1985–1994 the total capitalization of emerging markets rose by a factor of 11.2. There were 36,179 issues listed on all the world's exchanges at the end of 1996; some 17,115 companies of these were listed on emerging market's exchanges. This represents 47.3% of the total. Since 1985 the number of issues listed on emerging market's exchanges increased by 50%, mainly because of the large number of privatizations and initial public offerings. Over the same period the capitalization of the developed markets rose by a factor of 2. Some sources project that by the year 2025 the emerging markets' share of the total world stock market capitalization will rise from 12% to about 43% if the current growth rates of the emerging markets and those of developing countries are maintained.

In this chapter we apply self-organizing maps (SOM) to create two-dimensional maps on emerging stock market data. We provide first a brief overview of the performance of these emerging markets and compare return and risks in classical fashion. Next, we discuss the SOM approach for creating maps of emerging markets. Finally, we show how these maps can be used for asset allocations and decision-support and conclude with the strategic importance of maps for investing in emerging markets.

6.2 Performance and Risks of Investing in Emerging Markets

In the period 1989–1995 the IFC Investable Composite Index on Emerging Markets (with dividends reinvested) rose from $100 to $330 (18.5% per year). In contrast the Morgan Stanley Composite Index (MSCI) of all World Stocks rose from $100 to $169 (7.7% per year). The standard deviation of monthly returns of the IFC (Investable) Composite Index in the same period was 5.95%; the standard deviation of the MSCI index was 4.02%. The investment performance and potential of emerging markets rest in large part on their tremendously high growth rates.

The growth of the underlying economies is one of the most important determinants of the growth of the emerging markets and the earnings of companies

[6.02]. Keppler and Lechhner (1997) cite the following main factors that cause above-average growth rates of emerging markets:

- more favorable demographic fundamentals;
- rising levels of education and training;
- competitive labor costs, working hours and work morale;
- stability of political systems;
- upgrading of infrastructure;
- technology transfer;
- management of the debt crisis;
- improved currency stability;
- declining inflation rates;
- privatization;
- shrinking public sector;
- high savings rates;
- increased influx of foreign exchange;
- deregulation and liberalization of financial markets;
- increased world trade.

Investments in individual emerging markets carry also certain risks. The most common ones listed by Keppler and Lechhner are:

- political instability;
- high levels of foreign debt and currency risk;
- trade wars and trade restrictions;
- low and declining commodity prices;
- lopsided economic development;
- inadequate regulatory oversight of securities exchanges;
- lack of transparency and liquidity;
- broker, settlement and custodian risks.

In spite of all these risks, private investment flows towards emerging markets amounted in 1996 to approximately 230 billion US dollars. In recent years emerging markets have not only registered above-average performance, but their markets have also exhibited low correlation, both among themselves and with the developed markets. In the period 1989 to 1995 the coefficient of correlation between the MSCI World Index and the S&P 500 Index of US stocks was 0.69 (a high degree of positive correlation), whereas the correlation between the MSCI World Index and the IFCI Composite Index for emerging markets was only 0.44. There is a low degree of correlation between the Europe/Middle East/Africa regional index and the IFCI Composite index (0.44) and the other regional indices (Latin America 0.1 and Asia 0.27). More detailed tables on the correlation of the performance between various emerging markets and those of the developed world can be found in the *IFC Emerging Market Factbook*.

These are some highlights on the performance and risks of investments in emerging markets. Most books and magazine articles do not extend beyond simple comparisons of returns, standard deviations, correlations, or specialized ratios (e.g. the Sharp and Keppler ratios) for the emerging markets as a group. Some books dive from the aggregate, macro statistics into the micro, individual market-by-market descriptions.

6.3 Patterns Among Emerging Markets

The analysis in this section is based on multiple data dimensions of emerging markets and uses self-organizing maps to reduce the data to one- and two-dimensional maps. Instead of focusing on performance, risks or risk-adjusted returns we will map emerging markets based on returns, risks, market size, volume of transactions, price-earning ratios, price book values, dividend yield and other variables. First we discuss the data used (Section 6.3.1); next the methodology employed (Section 6.3.2); then the results from actual processing of the data (Section 6.3.3); finally (in Section 6.3.4), we present the detailed results from processing (i) time-series analyses, (ii) cross-sectional and (iii) geographical information of emerging markets.

6.3.1 Emerging Market Data

The Emerging Markets Database of IFC is a vital statistical resource used by the international financial community for investments in emerging markets [6.03]. In its second decade, this database has gained recognition as the world's premier source for reliable and comprehensive information and statistics on stock markets of developing countries. This database currently covers 44 markets, and includes updates on almost 2000 stocks comprising the IFC Global Composite Index. Out of these some 1200 stocks comprise the IFC Investable Composite (IFCI). The IFCI adjusts the Global Composite to reflect the accessibility of markets and of individual stocks for investors.

6.3.2 SOM as a Non-linear 2D Regression Model

Most neural network techniques are multi-layer, non-linear regression techniques whereby a given set of inputs are regressed against one or more outputs (Deboeck, 1994; Gia-Shuh Jang, 1994). Likewise, self-organizing map (SOM) is a non-parametric regression technique which is in most cases used to form a two-dimensional output (but could also be used for three-dimensional representations). SOM is a data-driven approach to extract relationships from data. The neurons of a SOM represent the general form of the data and quantize the input space. Training of neurons of a SOM through successive presentation of sample data vectors produces an "elastic net" that is stretched to cover the input space of the data. Instead of looking at vast quantities of tabular data, a SOM provides a map that allows to visualize an abstraction of the original data. Furthermore, a component plane representation of a SOM (see Chapter 3, Section 3.4.3.5) provides information about the correlations between data, the division of data in the input space, and relative distributions of the components.

In training a SOM it is very important to decide about how much data compression is desired, i.e. how much abstraction will be made of the original data. If a SOM is trained on a large number of neurons (which may require a lot of computational work) then one may end up with mapping small details. On the other hand, if a small number of neurons is chosen, then it may be impossible to differentiate between essential differences. *The optimal number of neurons or size of a SOM is therefore a question of selecting the optimal granularity or abstraction of the data for the purpose that the map will be used for.*

6.3.3 Computation of SOM Maps

All maps shown in this chapter were produced using the SOM_Package, which is public domain software developed by the Neural Network Center at the Helsinki University of Technology (HUT) (Kohonen et al., 1995). The features and capabilities of the SOM_Package are discussed in detail in Chapter 13.

Processing of most maps, including selection of "best" maps was undertaken by Antti J. Vainonen of the Neural Network Center.[1]

In preparing these maps we followed the following procedure:

- extraction of data from the IFC Emerging Market Database (using IFC software running under Windows 3.1);
- transfer of the data to an Excel spreadsheet for selection of appropriate data columns and rows;
- ftp of the selected data and pre-processing of the data at HUT (using "scale", a program developed at HUT to normalize the data);
- running of the SOM_Package at HUT;
- ftp transfer of the outputs from the SOM_Package to a desktop in Arlington;
- conversion and manipulation of the SOM outputs.

---

[1] Some maps were produced relying on concepts derived from "network-centric computing": through telnet and use of ftp individual programs of the SOM_Package were used from a desktop in the United States to create SOMs and produce visual maps on computers at the Neural Network Center of HUT. The main advantage of this was speed (the UNIX computers at HUT are a lot faster than a desktop), time savings (no time was needed for installation and maintenance of the software), greater platform independence (the SOM_Package is written to run under DOS and UNIX systems and conversion to other operating systems would be time-consuming) and efficient access (in addition to access to the latest release of the software, the timezone difference between Helsinki and Arlington provided, at least in the late afternoon and evening hours, access to underutilized night-time computing power in Helsinki).

Through an agreement reached with the Neural Network Center at HUT it was feasible to run the SOM_Package programs at HUT from a desktop in Arlington. This has allowed more time for analysis and interpretation of the results as well as better integration of SOM outputs into desktop publishable figures for this report.

The selection of "best" maps was in each case done on the basis of multiple simulations using different map initializations, different neighborhood functions, map topologies, and/or training procedures.

6.3.4 Main Results

The maps in this section were derived from (i) time-series analysis of 30 emerging stock markets, (ii) fundamental data on individual companies (banking, telecommunications and construction), and (iii) combined features of both emerging and frontier stock markets.

Table 6.1 contains basic data on 30 emerging markets as of the end of 1996.

Table 6.1. Data on 30 emerging markets as of the end of 1996

Country	Symbol	No. companies	Market capitalization (US$ mil)	Market dividend yield	Market P/E ratio	Market P/B ratio
Argentina	ARG	149	44,679.29	0.00	28.93	1.60
Brazil	BRA	550	216,989.81	0.00	0.00	0.67
Chile	CHI	291	65,940.2	0.22	13.35	0.73
China	CHA	540	113,754.58	0.00	37.25	0.00
Colombia	COL	189	17,137.15	4.55	0.00	0.00
Czech Rep.	CZR	1588	18,076.54	1.36	12.69	1.64
Egypt	EGY	646	14,172.63	6.28	12.45	0.00
Greece	GRE	224	24,178.18	4.76	11.55	2.57
Hungary	HUN	45	5,273.47	1.43	16.49	1.63
India	IND	5999	122,604.61	1.80	12.20	2.38
Indonesia	IDS	253	91,016.34	0.00	19.58	1.65
Jordan	JOR	98	4,551.27	2.38	17.01	1.75
Korea	KOR	760	138,817.25	1.40	16.10	0.86
Malaysia	MAL	621	307,178.78	1.30	28.56	4.72
Mexico	MEX	193	106,539.91	0.03	13.48	1.76
Morocco	MOR	47	8,704.92	2.45	18.50	2.57
Nigeria	NIG	183	3,559.91	6.05	6.97	0.00
Pakistan	PAK	782	10,638.75	0.00	0.00	0.00
Peru	PER	240	13,836.85	0.00	0.00	0.00
Philippines	PHI	216	80,648.62	0.83	26.32	3.98
Poland	POL	83	8,389.87	1.20	14.70	2.07
Portugal	POR	158	24,659.62	3.20	17.40	1.90
Russia	RUS	73	37,229.67	0.93	4.16	0.40
S. Africa	SAF	626	241,571.14	2.47	14.62	0.00
Sri Lanka	SRL	235	1,848.19	4.10	10.70	1.10
Taiwan, China	TWN	382	273,607.67	3.38	29.01	3.20
Thailand	THL	454	99,827.56	350	11.97	1.58
Turkey	TUR	229	30,020.06	2.87	12.15	4.54
Venezuela	VEN	87	10,055.01	0.13	12.83	1.98
Zimbabwe	ZIM	64	3,634.72	3.61	13.73	2.15

The main columns in this table show the number of companies listed, the market capitalization in millions of US dollars, the market dividend yield, and P/E and P/B ratios as of the end of 1996. After scaling this information so that the standard deviation of each column was set to one, we developed a SOM which produces a two-dimensional representation of this basic data. Figure 6.1 shows a 6 × 4 map of this data. The data vectors in Table 6.1 were spread over 6 × 4 or 24 neurons and vectors that are most similar were grouped together, while vectors that are most different appear further away from each other on the map. This map further shows a clustering into five zones:

- zone I: Argentina, Chile, Philippines, Malaysia and Taiwan (all to the right on the darkest neurons);
- zone II: Columbia, Zimbabwe, Portugal, South Africa, Thailand (in the center left of the map) and Mexico and China which are in between zones I and II;
- zone III: Egypt, Morocco, Russia, Hungary, Poland, Venezuela and Indonesia (all to the left on the lightest neurons);
- zone IV: includes India and Korea, which are in between zones III and V;
- zone V: Brazil, Czech Republic, Sri Lanka, Pakistan, Jordan, Peru, Greece, Nigeria and Turkey (all at the bottom of the map).

An analysis of the main differences between these five zones is shown in Table 6.2. The average market capitalization of the emerging markets identified by the SOM map varied from $154 billion (in zone I) to $24 billion (in zone III); the average P/E ratios from 25.2 (in zone I) to 11.8 (for the markets in zone V). Note that markets in zones I and II have above average P/Es, those in zones III and IV have close to average P/Es, and markets in zone V have below average P/Es. Only markets in zone I have above average P/B ratios; P/B ratios of markets in zones III and IV are below average, and P/B ratios of markets in zones II and V are about the same as the average P/B for all markets.

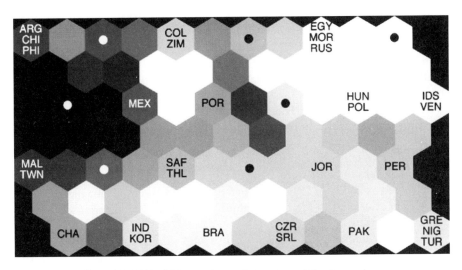

Figure 6.1. SOM of 30 emerging markets, using a 6 by 4 map size.

Table 6.2. Analysis of SOM zones of emerging markets as of the end of 1996

	No. of listed companies	Market capitalization (US$ mil)	Market dividend yield	Market P/E ratio	Market P/B ratio
Zone I (5 emerging markets)					
Average	332	154,411	1.4	25.2	2.8
Maximum	621	307,179	3.4	29.0	4.7
Minimum	149	44,679	0.2	13.4	0.7
Zone II (7 emerging markets)					
Average	318	86,732	2.9	18.1	1.8
Maximum	626	241,571	4.6	37.3	2.2
Minimum	64	3,635	0.0	12.0	1.6
Zone III (7 emerging markets)					
Average	176	24,977	2.1	14.1	1.5
Maximum	646	91,016	6.3	19.6	2.6
Minimum	45	5,273	0.1	4.2	0.0
Zone IV (2 emerging markets)					
Average	3380	130,711	1.6	14.2	1.6
Maximum	5999	138,817	1.8	16.1	2.4
Minimum	760	122,605	1.4	12.2	0.9
Zone V (9 emerging markets)					
Average	459	35,967	3.6	11.8	2.0
Maximum	1588	216,990	6.1	17.0	4.5
Minimum	98	1,848	1.4	7.0	0.7
All emerging markets					
Average	534	71,305	2.5	16.6	2.0
Maximum	5999	307,179	6.3	37.3	4.7
Minimum	45	1,848	0.0	4.2	0.0

6.3.4.1 Time-series Analysis

The SOM presented so far did not include any data concerning the performance of these markets. In this section we apply SOM to the market price indices of 30 emerging markets. Figure 6.2 shows a SOM based on 52 weekly market price indices for 1996. Figure 6.3 shows a SOM based on 36 monthly market price indices from the beginning of 1994 to the end of 1996. Summary statistics on the returns and volatility, as measured by the standard deviation of returns in each market, are shown in Table 6.3.

In comparing the map in Figure 6.2a with the statistics in Table 6.3, we notice that three markets which have very high volatility – as measured by the standard deviation of their returns in 1996 – (i.e. China, Russia and Venezuela) all cluster in the left-hand top corner of the map. The stock markets of Portugal, Morocco, Jordan and Taiwan, which are all in the center bottom part of the map, had relatively low volatility in 1996. The stock markets of Argentina, Brazil, Indonesia and Mexico, clustered in the right bottom part of the map, experienced relative moderate volatility in 1996.

The weekly market price indices for this SOM were preprocessed so that the columns are normalized to a standard deviation of one. If we compare the normalized series for China and Mexico, as shown in Figure 6.2b, the *relative*

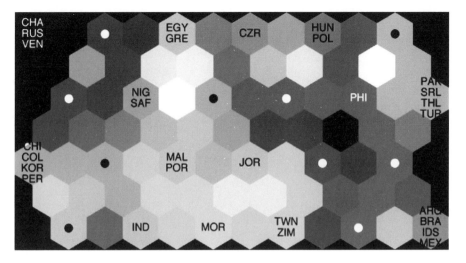

Figure 6.2a SOM of the weekly returns of 30 emerging markets, using a 6 by 4 map size.

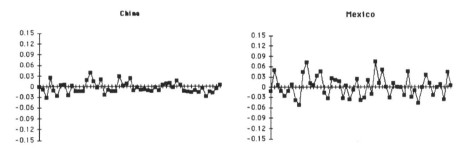

Figure 6.2b Volatility of China and Mexico market price indices relative to volatility of price indices of all emerging markets.

volatility of the Mexican market was actually much higher than the relative volatility of the Chinese market in 1996. Relative volatility here means volatility of a market relative to all 30 emerging markets. In consequence, the SOM map in Figure 6.2a actually clusters markets with low relative volatility in the top left corner and markets with high relative volatility in the bottom right corner.

A similar analysis was performed for 36 monthly returns from the beginning of 1994 to the end of 1996. Figure 6.3 shows that China and Turkey, which had high volatility as measured by standard deviation or their monthly returns (see Table 6.3), actually had low relative volatility as compared with all emerging markets and low relative volatility as compared with the markets clustered in the right bottom of the map, as shown in Figure 6.3b.

Also included on the map in Figure 6.3a is the EAFE index which produced 12% return in the period 1994 to 1996 and clusters in the middle of the map with Taiwan

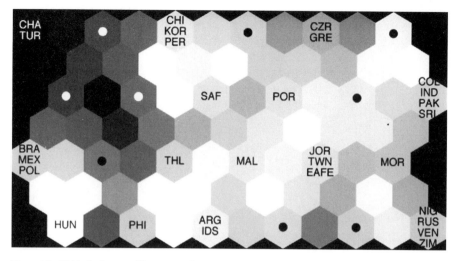

Figure 6.3a SOM of the monthly returns from 1994 to 1996 of 30 emerging markets, using a 6 by 4 map size.

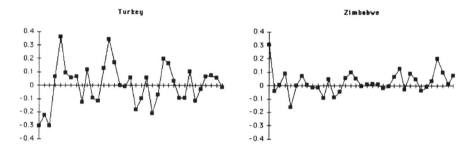

Figure 6.3b Volatility of Turkey and Zimbabwe market price indices relative to volatility of price indices of all emerging markets.

and Jordan. The average monthly return on the EAFE in 1994–1996 was 0.37% and the standard deviation of the monthly returns was 2.9%.

The previous analyses produced maps that cluster markets based on relative volatility either in a single year or over a period of years. SOM can be applied to obtain a more comprehensive view of the evolution of emerging markets. As an illustration we show in Figure 6.4 four one-dimensional SOMs based on multi-dimensional input data sets taken at the end of 1988, 1990, 1993 and 1996.

From this figure we note that in 1988 the emerging markets cluster in three groups: Korea and Taiwan in group I on the left of the map, Portugal and Philippines in group II on the right of the map, and all other countries in the middle of the map. By 1990 four groups can be distinguished with Indonesia, India and Malaysia joining group I, and Chile. Mexico, Argentina and Thailand joining group II. By 1993 five groups can be distinguished. By the end of 1996 we note

Table 6.3. Summary of returns and volatility in emerging markets sorted by highest volatility for 1996 and for 1994 to 1996

Period 1996					Period 1994–1996				
Country	Sym	Total return	Average weekly returns	Std. dev. weekly returns	Country	Sym	Total return	Average monthly returns	Std. dev. monthly returns
Russia	RUS	150.0%	2.31%	10.29%	China	CHA	25.1%	2.80%	26.21%
Venezuela	VEN	140.1%	1.92%	6.39%	Russia	RUS	150.0%	9.72%	21.61%
China	CHA	69.1%	1.23%	6.30%	Poland	POL	-35.6%	0.02%	16.19%
Turkey	TUR	32.4%	0.66%	4.71%	Venezuela	VEN	50.3%	2.52%	15.96%
Poland	POL	50.8%	0.89%	4.12%	Turkey	TUR	-23.3%	0.43%	15.52%
Pakistan	PAK	-19.8%	-0.37%	3.56%	Nigeria	NIG	17.7%	2.14%	13.39%
Brazil	BRA	39.6%	0.71%	3.42%	Mexico	MEX	-52.3%	-1.30%	12.09%
India	IND	-1.8%	0.02%	3.32%	Brazil	BRA	52.6%	1.86%	11.95%
Hungary	HUN	105.1%	1.47%	3.31%	Hungary	HUN	30.0%	1.18%	9.50%
Mexico	MEX	6.5%	0.17%	3.11%	Peru	PER	8.0%	0.60%	9.15%
Argentina	ARG	5.0%	0.14%	3.04%	Argentina	ARG	-20.4%	-0.25%	8.95%
Thailand	THL	-39.7%	-0.94%	2.97%	Zimbabwe	ZIM	134.9%	2.79%	8.36%
Korea	KOR	-29.4%	-0.64%	2.93%	Taiwan	TWN	8.9%	0.53%	7.81%
Columbia	COL	6.3%	0.16%	2.91%	Sri Lanka	SRL	-54.8%	-1.95%	7.78%
Zimbabwe	ZIM	88.8%	1.29%	2.81%	Thailand	THL	-44.7%	-1.42%	7.25%
Egypt	EGY	38.9%	0.68%	2.63%	India	IND	-46.3%	-1.51%	7.21%
Indonesia	IDS	17.4%	0.35%	2.62%	Czech Rep.	CZR	-46.1%	-1.49%	7.14%
Peru	PER	-2.2%	-0.01%	2.59%	Pakistan	PAK	-54.6%	-1.98%	7.07%
Philippines	PHI	17.2%	0.34%	2.56%	Indonesia	IDS	-3.6%	0.12%	6.83%
Taiwan	TWN	42.1%	0.72%	2.50%	Columbia	COL	-21.5%	-0.48%	6.58%
Czech Rep.	CZR	18.7%	0.36%	2.09%	Philippines	PHI	14.1%	0.58%	6.53%
Sri Lanka	SRL	-14.2%	-0.28%	2.06%	Korea	KOR	-34.2%	-1.01%	5.98%
Greece	GRE	-1.3%	-0.01%	2.02%	Chile	CHI	4.7%	0.30%	5.86%
South Africa	SAF	-22.0%	-0.47%	1.90%	Malaysia	MAL	22.6%	0.75%	5.83%
Chile	CHI	-18.7%	-0.39%	1.53%	S. Africa	SAF	31.9%	0.95%	5.66%
Jordan	JOR	-2.8%	-0.04%	1.50%	Greece	GRE	-10.8%	-0.22%	4.57%
Malaysia	MAL	17.1%	0.32%	1.38%	Jordan	JOR	-10.3%	-0.24%	3.64%
Morocco	MOR	29.6%	0.52%	1.36%	Portugal	POR	20.8%	0.60%	3.32%
Portugal	POR	22.4%	0.40%	1.14%	Morocco	MOR	28.4%	2.16%	3.27%
Nigeria	NIG	50.6%	0.81%	1.13%	EAFE	EAFE	12.0%	0.37%	2.94%

clustering on each neuron going from Chile, Columbia and Philippines on the left to Hungary, Russia, Indonesia and Morocco on the right of the map. As markets evolve they assume different patterns that cluster in different ways and jump from cluster to cluster.

To gain a better insight on how emerging markets evolve over time we merged all data sets from 1988, 1990, 1993 and 1996 and created a single SOM whereby multi-dimensional input data vectors of each market and for each selected year appears with its respective country label and year. The result is the SOM map shown in Figure 6.5.

The arrows in Figure 6.5 show the evolution of individual markets over time. For example, Korea which was part of group I in 1988 moves up and to the right in 1990

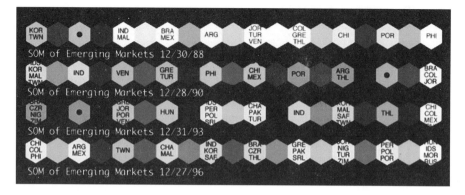

Figure 6.4. SOM of emerging markets at the end of 1988, 1990, 1993 and 1996, using a 1 by 10 map size.

and 1993 and then jumps to the top right corner of the map in 1996. On the other hand, Chile starts off from the top and falls to the right bottom of the map by 1996. Almost all markets seem to move from left to right on the map. Two interesting exceptions are Brazil and Turkey: Brazil starts at the top of the map, moves to the bottom left and then up to the right top; Turkey moves zigzag until in 1993.

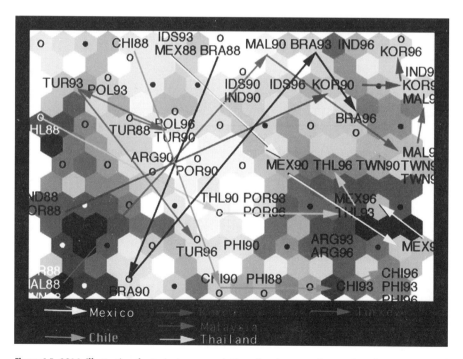

Figure 6.5. SOM illustrating the trajectory or evolution of major emerging markets from 1988 to end of 1996.

6.3.4.2 Cross-sectional Analysis of Emerging Market Companies by Sector

Through the creation of maps of fundamental or technical data, it is possible to detect similarities and dissimilarities among markets and trace their evolution over long periods of time. We moved beyond classical analysis of emerging markets by creating maps using multi-dimensional input data sets as well as by mapping relative volatility among markets as opposed to absolute volatility. In this section we will apply SOM to micro data: individual company data of corporations belonging to the same sector across the globe. We used the IFC's emerging market company database, which consists of data on over 1600 companies. We will focus on companies in the banking, telecommunications and construction sectors.

A SOM of 144 companies in banking in emerging markets is shown in Figure 6.6 [6.04]. If we focus on just one attribute, the P/E ratio of these individual banks, then we note that the banks shown in the top left-hand corner of the map (Akbank and T. Is bank (C)-Beaver in Turkey) have P/E ratios of 9.1 and 9.07 respectively. At the other extreme of the map, in the right-hand bottom part, the Commercial Bank in Greece, the Banespa Bank in Brazil, the First Bank of Korea and the Tong Yang Merchants Bank in Korea have P/E ratios between 0.34 and 5.1. At the top right-hand corner of the map we find the Bank of Seoul and the Hanwha Merchant Bank with P/E ratios of 139.4 and 79.6 respectively. This SOM on banks in emerging markets also takes into account the number of shares outstanding, the market capitalization, the number of shares traded, the P/B ratio and the dividends paid in 1996. Looking at a single attribute does not do justice to the clustering presented in this map.

A SOM of 48 individual companies in the telecommunications sector in emerging markets is shown in Figure 6.7. Again, if we were to focus on just one

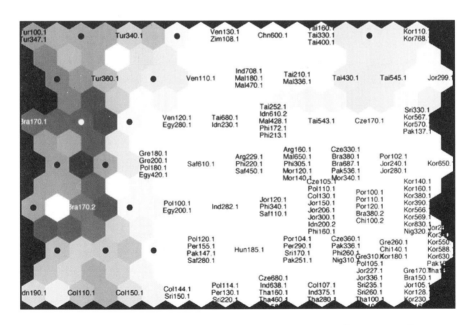

Figure 6.6. SOM of 144 individual companies in the banking sector of emerging markets.

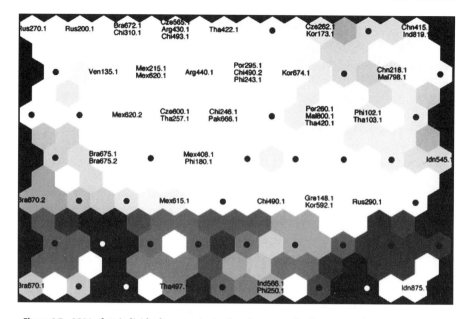

Figure 6.7. SOM of 48 individual companies in the telecommunication sector of emerging markets.

attribute, the P/E ratio of these individual companies, then we note that one of the companies shown in the top right-hand corner of the map, Oriental Pearl of China, has a P/E of 56.4, Moscow Telephone in the top left corner has a P/E of 23.8, while Telebras of Brazil in the bottom left corner has a P/E of 9.2. In the

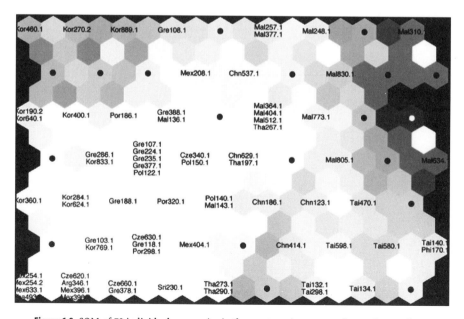

Figure 6.8. SOM of 72 individual companies in the construction sector of emerging markets.

bottom row P/E's move from 9.2 for Telebras in Brazil, to 19.5 for MTNL in India, to 32.9 for Telkom in Indonesia.

A SOM of 72 companies in the construction sector in emerging markets is shown in Figure 6.8. Once more, it is important to underline that these SOMs do not merely provide a rank-ordering of companies by P/E ratios, but take into account seven fundamental data attributes including company size, shares and value traded (in 1996), P/E and P/B ratios as well as dividends paid.

6.3.4.3 Geographic Analysis of Emerging and Frontier Markets

The same analyses as in previous sections can also be conducted on 14 frontier markets, i.e. markets which are even newer or younger than the emerging markets and on which there is less data.

A list of the major frontier markets can be found in Table 6.4. The number of stocks listed and the market capitalization of these markets is small . The average number of stocks listed in frontier markets is 128 as compared with 533 in emerging markets. The market capitalization of frontier markets varies from as little as $7 million in Bulgaria to $4 billion dollars in Tunisia. Table 6.5 highlights some other differences: average market capitalization is $1.7 billion in frontier markets as compared with $35.7 billion for emerging markets; the average P/E of the frontier markets in 1996 was 13.1 as compared with 14.4 for emerging markets. The range of P/E ratios is much wider: the range for frontier markets varies between 7.3 to 69, while the maximum P/E among emerging markets was only 38.4 in 1996.

Figure 6.9 provides two maps: a one-dimensional map that puts Bangladesh on one end of the map and the two 'S' countries (Slovakia and Slovenia) on the other end of the map; a two-dimensional SOM that spreads 14 frontier markets over a 6 × 4 grid. The input variables used for the training of these two maps included: the number of companies listed, the market capitalization in US$ millions, the volume

Table 6.4. Selected frontier markets

Country	Symbol	No. stocks	Market cap. (US$ mil)	Dividend yield	P/E	P/BV
Bangladesh	BANG	186	4551	10.1	69	4.1
Botswana	BTSW	12	325.53	7.5	7.28	2.29
Bulgaria	BUL	15	7.32	0.28	13.82	0.99
Cote d'Ivoire	CDI	31	914.44	NA	NA	NA
Ecuador	ECUD	42	1945.5	7.97	10.84	2.14
Ghana	GHA	21	1492.46	7.05	7.95	1.62
Jamaica	JAM	46	1887.27	2.49	7.93	1.13
Kenya	KEN	56	1846.18	5.4	15.13	1.95
Lithuania	LIT	460	900.27	NA	NA	NA
Mauritius	MAU	40	1676.3	3.97	11.37	NA
Slovakia	SLVA	816	2182.32	2.8	8.3	0.5
Slovenia	SLVE	21	663.16	NA	NA	NA
Trin.-Tobago	TRTO	23	1404.57	4.13	10.9	0
Tunisia	TUN	30	4263.36	0	20.4	2.9

Table 6.5. Comparison of main features of frontier markets with emerging markets

	No. stocks	Market cap. (US$ mil)	Dividend yield	P/E	P/BV
Frontier markets (14)					
Average	128	1,718.5	10.1	13.1	1.2
Maximum	816	4,551.0	8.0	69.0	4.1
Minimum	12	7.3	0.3	7.3	0.5
Emerging markets (30)					
Average	533	71,304.7	13.5	14.4	1.5
Maximum	180	307,178.0	6.3	38.4	4.0
Minimum	13	1,848.2	0.0	6.2	0.4

traded in US$ millions, the turnover rate, the number of shares traded, the dividend yield, the P/E ratio and the Price / Book value of the market.

We leave it up to the reader to interpret these self-organizing maps of frontier markets. Clearly, the scarcity of data on some markets plays a role; nevertheless it is interesting to observe how even partial data can be used by SOMs to create maps.

Figure 6.10 shows a geographical presentation of a SOM based on both the emerging market data discussed in Sections 6.3.4.1 and 6.3.4.2 and the frontier

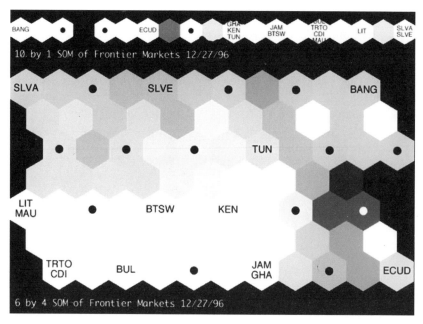

Figure 6.9. SOM of frontier markets among the developing countries. The first SOM (1 by 10 map) was obtained through an initial training session of 1000 cycles, started with a learning rate of 0.1 and a starting radius of 3; the second phase training was for 10,000 cycles, with learning rate 0.02 and radius 1. The second SOM is a 6 by 4 map obtained through an initial training session of 10,000 cycles, started with a learning rate of 0.1 and a starting radius of 3; the second phase training was for 100,000 cycles, with learning rate 0.02 and radius 1.

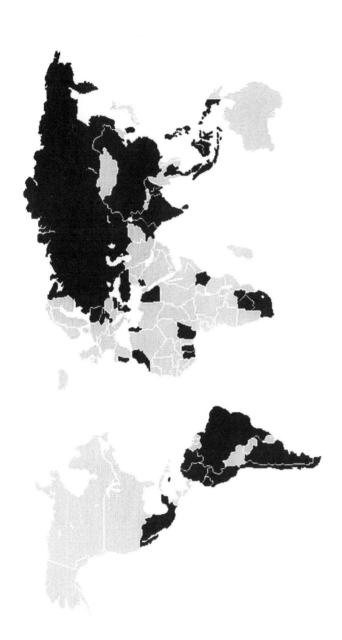

Figure 6.10. Self-organizing map of emerging and frontier equity markets grouped into five clusters, including:

- group 1: China, with a very high P/E of 30, high P/B ratio, low dividend payments, and moderate-sized companies ($189 million capitalization per company);
- group 2: Czech Republic, Slovakia and India, with small P/Es and low P/B ratios, and a large number of very small companies ($47 million capitalization per company) on average;
- group 3: eight countries, including Indonesia and Malaysia, with average P/E ratios of 21 and high P/B ratios, and low dividend payments on average;
- group 4: 11 countries, mostly in Latin America but also including Russia, moderate-sized companies ($160 million capitalization per company), low P/E (12), very low P/Bs and moderate dividend payments on average;
- group 5: 21 countries, mostly in Africa, with P/E ratios of 14, high P/B ratios, and relatively high dividend payments.

(*This figure can be seen in color in the Color Plate Section.*)

markets discussed in Section 6.3.4.3. A single SOM was created from a data table in which emerging and frontier market data for 1996 was merged. Since the input signals in this case related to spatial or geographic data, it is possible to provide a compact representation of the selected data by superimposing a "smooth" color palette on the neurons of the SOM. A color code is thereby associated with each of the model vectors. The resulting global map summarizes nicely the similarities and dissimilarities among emerging and frontier countries.

Figure 6.10 shows five zones among the emerging and frontier markets of the world. Each zone comprises from one to 21 countries. An analysis if the main differences between these five groups of equity markets is summarized in Table 6.6. In short:

- **Group I is a market zone with moderate-sized companies with high P/E ratio, high P/B ratio, and low dividend payments (1%).** It includes China, with over 450 companies listed, total market capitalization of $85.9 billion (or approximately $189 million per company), a P/E ratio of 30.2 and a P/B ratio of 3.4 (as of September 1996).

- **Group II is a market zone with primarily small companies, with small P/E ratios and low P/B ratios.** The companies in this zone pay average dividends (2%). This zone covers three countries: the Czech Republic, Slovakia, and India. The equity markets of these countries have on average 2801 companies listed, market capitalization of $47.6 billion (or approximately $17 million per company), average P/E ratio of 11.1 and average P/B ratio of 1.5.

- **Group III is a market zone with large-sized companies with high P/E ratios, high P/B ratios, and low dividend payments (1.3%).** Belonging to this zone are Pakistan, Bangladesh, Malaysia, Indonesia, and Taiwan in Asia; Hungary in Europe; and Chile in Latin America. Among these markets the average number of stocks listed was 323, the average market capitalization $94.8 billion (or approximately $293 million per company), the average P/E ratio 21.4 and the average P/B ratio 2.67.

- **Group IV is a market zone with moderate-sized companies, low P/E ratios, very low P/B ratios and moderate dividend payments (2.5%).** Included are Mexico, Peru, Argentina, Columbia, Trinidad and Tobago, and Jamaica in Latin America; Korea in Asia; Russia and Bulgaria in Europe; and Egypt and Cote d'Ivoire in Africa. Average number of stocks listed on these markets in 1996 was 215, average market capitalization $34.2 billion (or approximately $160 million per company), average P/E ratio 12 and P/B ratio 0.96.

- **Group V is the largest market zone. It has primarily moderate-sized companies with average P/E ratios, high P/B value, and high dividend payments (3.8%).** Markets in this zone are Brazil, Venezuela and Ecuador in Latin America; Philippines, Thailand, and Sri Lanka in Asia; Portugal, Greece, Turkey, Lithuania and Jordan in Europe; and Nigeria, Botswana, Ghana, Mauritania, Zimbabwe, South Africa, Kenya and Tunisia in Africa. Average number of stocks listed in these markets was 186 in 1996, average market capitalization was $36.7 billion, average P/E ratio 14 and average P/B ratio 2.2.

Table 6.6. Analysis of five major groups among emerging and frontier markets

Group	Count	Average no. companies	Average capitalization	Average dividends	Average P/E	Average P/B
1	1	454	85,990	1.00	30.20	3.40[a]
2	3	2801	47,621	1.99	11.06	1.51
3	8	323	94,859	1.27	21.40	2.67
4	11	215	34,239	2.51	11.97	0.96
5	21	186	36,718	3.88	13.76	2.22

[a] China figures as of September 1996.

6.4 Strategic Implications of SOM for Investments in Emerging Markets

SOM can be deployed to reduce multi-dimensional data on emerging markets to two-dimensional maps. We thus demonstrate the strategic importance of using SOM for investing in emerging markets.

The neural networks outlined in this chapter can be used in a variety of ways for decision-support and asset allocation between markets. *Asset allocation is the distribution of investment capital among various asset classes such as stocks, bonds and cash; or between small company stocks, mid-cap and large company stocks; or between markets.* We will consider the implications of alternative capital allocations between the US stock market and emerging markets or between world stocks and emerging markets. Subsequently we will offer alternative investment strategies for emerging markets and show the impact of using SOM groupings for investing among emerging markets

For the period 1989 through 1995 the S&P 500 Index had a compounded annual return of 15.4%; the total return of the IFCI Composite Index was 18.6% annually (20.8% higher than the S&P 500). However, the risk as measured by the expectation of monthly loss was 1.48% for the IFCI Composite Index (72.8% higher than the S&P 500 expectation of monthly loss). Based on the risk-and-return characteristics of the US stock market and the emerging market equities, observed over the past seven years, Keppler and Lechhner recommends the following asset allocations for various classes of investors:

- *ultra conservative investors* should invest 90% in US stocks and 10% in emerging markets, which on a risk-return basis would have produced 15.9% per year with a Keppler ratio of 1.706;

- *conservative investors* should invest 80% in US stocks and 20% in emerging markets, which on a risk-return basis would have produced 16.4% per year with a Keppler ratio of 1.73;

- *adventurous investors* should invest 70% in US stocks and 30% in emerging markets, which on a risk-return basis would have produced 16.8% per year with a Keppler ratio of 1.7.

In comparison a 100% allocation to the US stock market only would have produced 15.4% return per year accompanied with a Keppler ratio of 1.55.

Based on the risk-and-return characteristics of the MSCI World Stock Index and emerging market equities observed over the past seven years Keppler et al. recommends the following asset allocations between the MSCI World Stock Index and the emerging markets be applied by various classes of investors:

- *ultra conservative investors* should invest 90% in MSCI World stock index and 10% in emerging markets, which on a risk-return basis would have produced 8.96% per year with a Keppler ratio of 0.68;

- *conservative investors* should invest 80% in MSCI World stock index and 20% in emerging markets, which on a risk-return basis would have produced 10.2% per year with a Keppler ratio of 0.79;

- *adventurous investors* should invest 70% in MSCI World stock index and 30% in emerging markets, which on a risk-return basis would have produced 11.3% per year with a Keppler ratio of 0.9.

In comparison a 100% allocation to the MSCI World stock index only would have produce 7.7% return per year accompanied with a Keppler ratio of 0.55.

Given the above asset allocations between the US, world stock and emerging markets, we should next consider asset allocation strategies and in particular focus on asset allocation strategies for emerging markets. Several strategies for organizing a portfolio on emerging market equities are outlined in Keppler et al. The most popular ones are:

- *Market-capitalization weighting.* Market-capitalization weighting of a global equity portfolio implies that the largest dollar amounts are invested in the largest markets. In the emerging equity markets five markets account for 71.4% of the total market capitalization: South Africa, Malaysia, Brazil, Mexico and Thailand. The main disadvantages of market-capitalization weighting are (i) the dynamic changes in the structure of the markets; (ii) a market-capitalization-weighted portfolio will not be a lowest cost portfolio.

- *Equal weighting.* An equal weighting of a global equity portfolio implies that each market has the same weight, with monthly, quarterly or annual rebalancing. An equally weighted portfolio has a higher expected rate of return than a market-capitalization-weighted index. Over the seven-year period from 1989 to 1995, the market-capitalization-weighted IFCI Composite Index had a total compounded annual return of 18.57% in US dollars. An equally weighted portfolio would have yielded 35.57% compounded annually.

- *Small market bias.* During the seven-year period from 1989 to 1995 the largest emerging markets as measured by their investable market capitalization had a total compounded annual return of 10.7% in US dollars, the medium-sized emerging markets returned 39.7% on average, and the small emerging markets returned 55.5% compounded per annum. This means that an equally weighted portfolio of the smallest markets outperformed the capitalization-weighted IFCI Composite Index by 37 percentage points per annum and outperformed the equally weighted IFCI Composite Index by 20 percentage points per annum.

- *Value market selection.* Certain global equity portfolios selected on the basis of value can be expected to have superior risk-return characteristics compared with a market-capitalization-weighted world index. Independent studies on global investing have shown that the selection of the market to invest in is more important than the stock selection decision. Global equity investors can achieve excess risk-adjusted returns over the long term by investing in markets with above-average dividend yields. A global equity portfolio selected on the basis of value would have yielded 60.3% compounded annually.

The above four strategies for asset allocation between various markets all rely on a *single criterion* (e.g. market-capitalization, equal weighting, small market size or value). In this chapter we have shown how SOM can reduce multiple dimension data to two dimensions which results in maps that cluster groups of markets with similar features. SOM maps can thus be used as a basis for asset allocations between distinct groups of markets.

As an illustration, let us assume four different weightings, as shown in Table 6.7. The second column in this table proposes to weight each SOM group equally, thus expressing no bias toward any of the main features or combination of attributes that a SOM may produce. The second set of weights in Table 6.7 proposes to give higher weight to the group with the highest dividends (which in Table 6.6 were groups 5, 4 and 2 (in order) while groups 3 and 1 had the lowest dividend payments). Since we propose higher weighting for the higher dividend group we shall start with 30% and go down to 10% for the group with the lowest dividend weighting. The third set of proposals in Table 6.7 propose to do the same for P/E groups. Table 6.6 showed that groups 2, 4 and 5 had average P/E ratio (in the order of 11 to 14); in consequence each of these groups is allocated one-third weighting, while the other two groups are left with zero weighting. The final column in Table 6.7 proposes that lowest price book groups get one-quarter of the weighting and that the highest price book group be allocated zero weighting. Groups 2, 3, 4 and 5 had P/B ratios of 0.95 to 2.6 in Table 6.6 and were thus given 25% weighting each. Note that there are alternative ways to set these weighting factors. It would be very interesting, for example, to apply genetic algorithms for determining the optimal weightings based on clustering of market features over time (Wallace, 1996). For now, we will stick to this relatively arbitrary way of setting group weightings.

Applying these various weighting assumptions to the SOM-based classification of markets demonstrated in Table 6.6 and the market return achieved in 1996 gives the results shown in Table 6.8. To simplify these results all frontier markets were removed and weights were applied only to the emerging markets.

Table 6.7. SOM weightings for investments in emerging markets

SOM group	Equal/Group	Highest div	Lowest P/E	Lowest P/B
1	0.20	0.10	0.00	0.00
2	0.20	0.20	0.33	0.25
3	0.20	0.15	0.00	0.25
4	0.20	0.25	0.33	0.25
5	0.20	0.30	0.33	0.25

Table 6.8. SOM-based weighting for investments in emerging markets in 1996

SOM group	Symbol	Return 96	Cap weighted	Equal weighted	Small weighted	SOM Weighted			
						Equal/ Group	Highest div	Lowest P/E	Lowest P/B
4	ARG	5.0%	0.1%	0.2%	0.3%	1.0%	1.3%	1.7%	1.3%
5	BRA	39.6%	4.0%	1.3%	0.0%	7.9%	11.9%	13.2%	9.9%
1	CHA	69.1%	3.7%	2.3%	0.0%	13.8%	6.9%	0.0%	0.0%
3	CHI	-18.7%	-0.6%	-0.6%	-1.7%	-3.7%	-5.6%	-6.2%	-4.7%
4	COL	6.3%	0.1%	0.2%	0.1%	1.3%	1.6%	2.1%	1.6%
2	CZR	18.7%	0.2%	0.6%	0.5%	3.7%	3.7%	6.2%	4.7%
4	EGY	38.9%	0.3%	1.3%	0.8%	7.8%	9.7%	13.0%	9.7%
5	GRE	-1.3%	0.0%	0.0%	0.0%	-0.3%	-0.2%	0.0%	-0.3%
3	HUN	105.1%	0.3%	3.5%	0.8%	21.0%	15.8%	0.0%	26.3%
3	IDS	17.4%	0.7%	0.6%	2.2%	3.5%	5.2%	5.8%	4.3%
2	IND	-1.8%	-0.1%	-0.1%	0.0%	-0.4%	-0.4%	-0.6%	-0.5%
5	JOR	-2.8%	0.0%	-0.1%	0.0%	-0.6%	-0.8%	-0.9%	-0.7%
4	KOR	-29.4%	-1.9%	-1.0%	0.0%	-5.9%	-4.4%	0.0%	-7.3%
3	MAL	17.1%	2.5%	0.6%	0.0%	3.4%	5.1%	5.7%	4.3%
4	MEX	6.5%	0.3%	0.2%	1.0%	1.3%	1.6%	2.2%	1.6%
5	MOR	29.6%	0.1%	1.0%	0.4%	5.9%	8.9%	9.9%	7.4%
5	NIG	50.6%	0.1%	1.7%	0.2%	10.1%	7.6%	0.0%	12.6%
3	PAK	-19.8%	-0.1%	-0.7%	-0.3%	-4.0%	-5.9%	-6.6%	-5.0%
4	PER	-2.2%	0.0%	-0.1%	0.0%	-0.4%	-0.5%	-0.7%	-0.5%
5	PHI	17.2%	0.6%	0.6%	1.9%	3.4%	5.2%	5.7%	4.3%
5	POL	50.8%	0.2%	1.7%	0.6%	10.2%	15.2%	16.9%	12.7%
5	POR	22.4%	0.3%	0.7%	0.8%	4.5%	6.7%	7.5%	5.6%
4	RUS	150.0%	2.6%	5.0%	7.7%	30.0%	37.5%	50.0%	37.5%
5	SAF	-22.0%	-2.5%	-0.7%	0.0%	-4.4%	-6.6%	-7.3%	-5.5%
5	SRL	-14.2%	0.0%	-0.5%	0.0%	-2.8%	-4.3%	-4.7%	-3.6%
5	THL	-39.7%	-1.9%	-1.3%	-5.5%	-7.9%	-11.9%	-13.2%	-9.9%
5	TUR	32.4%	0.5%	1.1%	1.3%	6.5%	4.9%	0.0%	8.1%
3	TWN	42.1%	5.4%	1.4%	0.0%	8.4%	12.6%	14.0%	10.5%
5	VEN	140.1%	0.7%	4.7%	1.9%	28.0%	42.0%	46.7%	35.0%
5	ZIM	88.8%	0.2%	3.0%	0.4%	17.8%	26.6%	29.6%	22.2%
Total			15.5%	26.5%	13.3%	30.1%	26.7%	21.5%	20.3%

Table 6.8 shows that if a market-capitalization weighting had been applied to the emerging markets (without rebalancing) the return from investments in emerging markets in 1996 would have been 15.5% per year. An equal-weighted portfolio would have produced 26.5% in 1996; a portfolio biased towards the smallest markets (that is those with less than 5% of the market capitalization of all emerging markets) would have produced 13.3%. In contrast if SOM-based weightings are applied we find that

- the equal based weighting between all five groups created by the SOM would have produced 30.1% in 1996;

- weighting of the markets based on SOM groups with the highest dividends (as shown in Table 6.7) would have produced 26.7% in 1996;

- weighting of the markets based on SOM groups with lowest P/E would have produced 21.5% in 1996;

- weighting of the markets based on SOM groups with lowest P/B would have produced 20.3% in 1996.

Given that SOM is a tool that reduces dimensionality of the data and that markets that cluster together have common features, it is not surprising that the returns of a portfolio based on SOM mapping are higher than those of any of the portfolios based on a single criterion.

SOM mapping is thus of strategic importance in asset allocations. As demonstrated above it can improve the return of a portfolio over any of the classic asset allocation strategies.

SOM is also an approach that can significantly improve on the construction of benchmarks and reduce the costs of maintaining and updating benchmarks. This should be particularly relevant to several large investment banks and international organizations who publish one or more composites of country indices.

6.5 Conclusions

Studies have shown that, regardless of the investor's base market and currency, global equity portfolios, over longer periods, offer higher returns at lower risk than investments in national markets only. The 30 emerging market economies (on which IFC publishes data) contain 4.8 billion people or 84% of the world's population. Of these 4.8 billion people, 2.7 billion live in the rapid growing economies of Latin America, China, India and Indonesia. In comparison, the developing countries contain 900 million people or 16% of the world's population. The investment potentials of emerging markets rest in large part on the tremendously high growth rates of emerging market economies. In a market that is rising but experiencing wide price swings, high volatility in the positive region of the return distribution means higher risk-adjusted return potential. Thus, higher risk-adjusted return should not be viewed as a risk to be avoided. In recent years emerging markets have not only registered above-average performance, as compared with the developed countries, but have also exhibited low correlation, both with each other and with the developed markets. On the assumption that the emerging markets will continue to grow, the emerging markets' share of the total world stock market capitalization will, according to Keppler et al., rise from 12.7% to about 43% by the year 2025.

In this chapter we have demonstrated how self-organizing maps can be created for reducing multi-dimensional data on emerging markets into one- or two-dimensional maps. These maps show similarities and dissimilarities between various markets and as a result it is possible to identify groups of countries with common features. The grouping of emerging markets is important for asset allocations between markets as well as for the creation of improved benchmarks and to reduce costs of maintaining an emerging market investment portfolio.

7 A Hybrid Neural Network System for Trading Financial Markets

Marina Resta

Abstract

In this chapter Marina Resta demonstrates experimentally the great potential of neural networks for design of systems for trading stock markets. She suggests the use of a hybrid neural network architecture that combines the approach of Self-Organizing Maps together with that of Genetic Algorithms. She shows the forecasting capabilities of this hybrid system and evidence of the performance of this system. This chapter is a nice extension of the applications of trading systems presented in "Trading on the Edge". The novelty here is that genetic algorithms and self-organizing maps are combined.

7.1 Introduction

Over the years forecasting of financial markets has been based on (i) rational expectations theory; (ii) time series analysis; and (iii) technical analysis. According to the rational expectations theory prices move up or down because investors react rationally and immediately to news: any differences between investors, say in investment objectives or horizons, are ignored as statistically irrelevant. This means that markets are completely transparent and no operator has information that others don't have. There is no competitive advantage. It's impossible to obtain more profit opportunities by receiving information that others do not have. Correlation between price movements in foreign exchange markets revealed strong contradictions with these theoretical assertions. Furthermore, market behavior during extreme events breaks down models at crucial moments (such as the stock market crash of 1987).

In *time series analysis* statistical methods are used to detect a limited number of determinants of price movements. This approach can pick out trends, however, serious difficulties arise when the series contains regularities or uniformly shaped cycles.

In *technical analysis* one tries to beat the market by charting techniques, and by combining for each security information on price, volume, and relative strength. Technical analysis methods rarely succeed in getting rid of the risk, or in constraining it within satisfactory boundaries. They are financial oriented systems; they are also strongly dependent on initial conditions. In sum, traditional forecasting techniques have achieved limited results in financial markets. They have limits: strong dependence on initial conditions; rigidity, i.e. the relationship among prices and their determinants varies over time and these changes can be abrupt; and limited explanatory power, i.e. traditional approaches attempt to give quantitative descriptions of qualitative factors or rules that require judgement.

Traditional techniques therefore need to be replaced by new systems which are more efficient in dealing with the market's structural instability.

An alternative approach comes from artificial neural networks. Approaches in this field are substantially different: neural networks are systems that can learn from experience and adapt to change in their environment. Neural techniques have the following advantages: non-linear modeling capabilities, robustness to noise, and generalization from examples.

A number of difficulties arise when neural networks are employed for economic data processing. In specific economical contexts (e.g. exchanges markets, market indexes), neural network-based simulations assume that inputs are related to the desired outputs or can be explained by a number of determinants. This means that we think that there is an "a priori law" governing the flow between the original time series and output set. This leads to problems in determining which are the representative patterns; which factors can be considered crucial determinants to represent input data; and what is a significant sample.

In these circumstances it is preferable to deploy Self-Organizing Maps (SOM), which are able to find structures and cluster patterns in situations in which outputs are not available. The drawback is that SOM have some structural rigidity themselves: because the map dimensions are defined upfront before training starts, furthermore neurons cannot be added during the training procedure.

These problems can be partially overcome by integrating a flexible neural structure. By developing a hybrid neural network model that combines SOM's learning procedure with an evolutionary model we created an Integrated Self-Organization and Genetics approach (ISOG). Combining SOM with genetic algorithms can offer improved quantitative and qualitative information about financial markets. Our discussion of this approach is organized as follows. Section 7.2 provides a general description of ISOG. We stress the substantial differences between ISOG and SOM. Section 7.3 shows the results of a comparison between SOM and ISOG performances, including the profitability of ISOG when employed for trading financial markets. We also prove that an ISOG-based system does better when used in market hedging strategies (even if operational costs have been put in). Finally, Section 7.4 provides conclusions and perspectives for future work.

7.2 ISOG: Integrated Self-Organization and Genetics

Our approach integrates SOM and genetics: the system retains the properties of SOM, but obtains a greater flexibility since we do not fix the map dimensions. The map dimensions evolve during learning. The process we use contains the following four steps.

- Step 1. Map initialization: the original structure consists of a kernel of neurons arranged into a planar lattice of small dimensions; its growth depends on evolutionary laws.
- Step 2. When an input sample is presented, each neuron of the map is evaluated according to SOM similarity criteria. In this way the procedure is able to find a winner, i.e. the neuron offering the best approximation to input.
- Step 3. New neurons are created and added to the map. The number of neurons belonging to each new generation depends on a growth factor, which is determined at the beginning of the procedure, and the spatial position in the map. The features of the new neurons are in part inherited from the winning neuron, and partly obtained from a list of random values within a prefixed range (which is similar to biological crossing over). The way the map will react to external stimulus depending on criteria can have an influence on:

$$map_t = f\left(map_{t-1}, reco, \alpha, threshold\right)$$

where:

- map_t is ISOG at time t;
- map_{t-1} is ISOG at time t-1;
- *reco* is a reduction coefficient. It indicates the way each neuron builds up its neighborhood, once it has been identified as a winner. The higher it is, the smaller the distance among neighboring neurons will be. From a mathematical point of view, it acts as a weight on the calculation of Euclidean distance among the leader and other neurons. *reco* will be higher in earlier steps, and reduce to lower values as the simulation goes on. A good solution is to choose a function properly controlling fluctuations in *reco*, rather than keeping it at constant values;
- α indicates the speed at which the amplitude of the leader neighborhood is reduced. This parameter influences the differentiation process, so that employing constant values rather than the evolution laws should get different final results;
- *threshold* is a constant.
- Step 4. The process starts over from step 1, until all inputs have been shown to the map.

The main differences between SOM and ISOG is obviously found in the kind of map the two algorithms generate. From a geometrical point of view, different features of ISOG and SOM can be identified in the representation of the input space they offer. The basic idea underlying ISOG architecture is to capture essentials, and to tune in on them, giving the map capabilities to "decide" in what direction it has to evolve. Hence, starting from an original small-sized map, ISOG will grow in a precise way, isolating input features it considers more promising. Map growth can be monitored, because new generations are indexed differently from preceding generations. In this way we maintain the two-dimensional structure of the map even if, in practice, neurons appear to be organized in different layers, according to the moment they have been generated. ISOG's final result appears to be quite different from the one we can expect by training a pure SOM.

7.3 Simulation Results

7.3.1 General Frame

Our simulation was run on both MIBTEL daily values and MIBTEL daily fluctuations from July 16, 1993 to July 4, 1997: the first 500 samples were employed as a learning set; the remaining values were used as a control set. MIBTEL (Milano Indice Borsa Telematico) is the official index of Milan Stock Exchange. We referred to daily closing values obtained from the EOS Database.

Inputs were organized in patterns of different amplitude, but here we shall refer only to the single case of 5-day moving windows. The key idea is that of scanning input data with a moving window of 5 days.

We applied these inputs to both a pure SOM and ISOG, taking different sized maps:

- in the case of SOMs we considered two-dimensional maps of the following sizes: 30×30, 25×25 and 20×20;
- in the case of ISOG we chose three different types of initial maps: 5×5, 7×7 and 10×10.

Simulation parameters were initially set at: $reco = 2$, $\alpha = 6$, and $threshold = 4$.

The choice of training the nets also on daily values can be explained as a first attempt to verify the capabilities of those tools in recognizing input shapes, even if they are influenced by trend factors. We think that if this condition holds, this result could be a good starting point for quantitative forecasting via neural networks.

The results of these simulations are shown in Figure 7.1. These results offer information about convergence properties of both SOM and ISOG nets. A first conclusion is that ISOG and SOM are powerful tools in pattern recognition, even when they are tested on financial time series data.

7.3.2 A Comparison of ISOG and SOM Performances: Discussion

To compare ISOG and SOM we used four metrics: convergence, scalability, generalization and stability.

Convergence properties have been already tested on daily values of the index, but results may be provided also by referring to MIBTEL daily fluctuations. Scalability is related to convergence properties in the sense that the greater the number of neurons in the map, the more time will be required to perform a single learning iteration. In a nutshell, network complexity affects the computing time.

The smallest sized ISOG net (initialized at 5×5 cells) takes on average 16 minutes to be trained over the whole input set. In contrast, the smallest SOM took

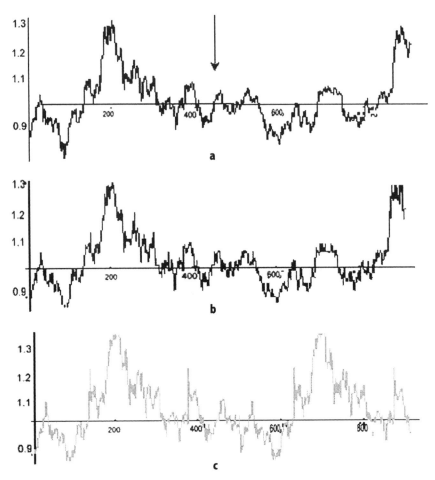

Figure 7.1. A comparison among **a** the real MIBTEL values, **b** the results from ISOG and **c** SOM. The arrow indicates the point until which the nets have been trained. To the right of the arrow are the neural net results based on control samples. Thus values on the right of the arrow have been presented to the nets as outside patterns. Both kinds of nets provide performances that are not influenced by trend phenomena.

33 minutes. The largest ISOGs require 31 minutes vs. 75 minutes to train the largest SOM. This difference is even more evident if we compare processing times for each cell: processing each neuron takes 3 to 4.2 seconds in ISOG; the same task requires from 4.5 to 6 seconds for SOM training. At the end of the process ISOGs has a minimum of 350 neurons (starting from a 5 × 5 map) and a maximum of 500 neurons (starting from a 10 × 10 map); in contrast the size of a SOM map is set at the beginning of the training procedure and cannot be modified during the learning phase.

When we use ISOG we deal initially with a small dimensional map whose size is increased step by step according to a procedure controlled by means of the growing factors (*reco* and *threshold*). In this way we manage a smaller number of neurons for the initial steps of the procedure, speeding up the process in pattern recognition.

Another important test is generalization capabilities. The ability of the nets to recognize patterns outside the training set has been verified by means of financial criteria. In particular, we used a decision-making tool which integrates forecasting provided by neural nets with trading. We entered all the information given by the nets into a decisor (Figure 7.2) in order to obtain the operative decisions, i.e. daily selling / buying strategies for the market, and to value their profitability. We made a number of hypotheses: (i) buying and selling are costless; (ii) the interest rate on standby positions was zero, which means that right responses are compensated at the current daily fluctuation rate, and wrong decisions are punished at the same (negative) rate, under the assumption of the compound interest rate rule.

The results are shown in Figure 7.3. From Figure 7.3 we see that the performances of both nets are satisfactory, although there are substantial differences: ISOGs provide results varying into a shorter range than SOM; losses in ISOG are bounded into a smaller range than the ones appearing with SOM.

These differences are confirmed when we take a look at stability properties of the two different kinds of map. Stability here means the ability of a net to get similar results each time the simulation is repeated with key parameters modified.

We repeated the simulation, changing maps initialization and key parameters to verify the way network behavior could resist different training conditions and test sets. In particular, we simulated entering the market at different times, acting as suggested by the neural decisor, and maintaining this position over short, middle or long terms.

Figures 7.4 to 7.6 show the stability tests we obtained from running SOM (a) and ISOG (b) net simulations. The maximum gains we achieved by entering the market and operating in the market for 25, 50 and 120 days are shown.

ISOGs and SOMs can handle different temporal windows (see Figure 7.7): the performance of SOM is represented by the first three bars, while ISOG's results are shown by the last three. Values offered by ISOG are always positive and greater than the ones provided by SOM.

When we consider a 25-day time frame, SOM gives results varying in the range from -0.02, to +0.098; ISOG produces results that range from +0.002 to +0.15. This

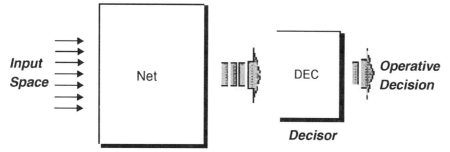

Figure 7.2. Patterns from the control set are sent to the ISOG and SOF. The response is then passed through a decisor (DEC) which provides a buy/sell recommendation for the market.

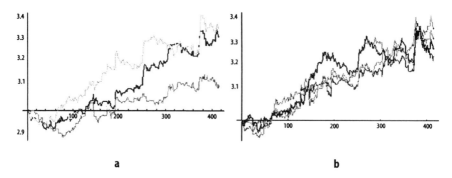

Figure 7.3. Results obtained from **a** SOM and **b** ISOG for uncovered patterns. Performances are both quite satisfactory even if ISOG shows more capabilities to maintain losses within more restricted boundaries.

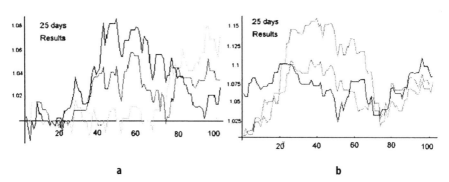

Figure 7.4. Results from **a** SOM and **b** ISOG, trading the market for 25 days.

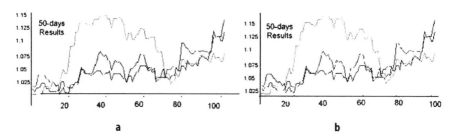

Figure 7.5. Results from **a** SOM and **b** ISOG, trading the market for 50 days.

situation is repeated when we turn to 50-day windows, except for the amplification in the better results of both nets, while the worst ones remain stable at the levels underlined before.

Finally, when we look at 120 days, ISOGs supremacy is confirmed: gains are in the range [+0.0765, +0.265], whilst SOM's results are in the range [+0.0001, +0.2]

Both these nets provide good performance in trading financial markets, but

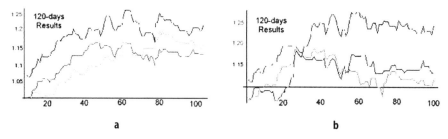

Figure 7.6. Results from **a** SOM and **b** ISOG, trading the market for 120 days.

considering the metrics we chose ISOGs produce better results than SOMs because they provide, on average, more stable and consistent results (and are computationally more efficient). One could, however, object that these results have been achieved assuming no transaction costs. In the next section, we prove that ISOGs guarantee satisfying profits, even when transaction costs are taken into account.

7.3.3 Arbitrage Opportunities by Joining ISOG-based Strategies

In this section we show the arbitrage opportunities which ISOG strategies can offer. A simulation was run under the following hypotheses:

- the ISOG nets were originally sized at 7×7 neurons;
- the nets have been trained over the first 830 MIBTEL daily fluctuations;
- the remaining 168 samples have been employed as a control set;
- ISOG capabilities to recognize turning points have been employed in identifying trading opportunities.

In addition, we assumed that:

- both successful and unsuccessful trades are compensated at the same rate (positive in case of gains, negative in case of losses),
- neutral positions are paid at zero rate;
- it was possible to invest in the market by borrowing money at an annual rate $[i]$;
- we have taken the transaction costs $[c]$ into account each time we entered or came out of the market.

The latter conditions have been introduced to make ISOG performances comparable to the ones provided by operating random decisions and acting as insiders in the market. Our goal was to find the combination of passive interest rate and transaction costs that operators should accept to avoid losses. We chose three different windows: ISOG arbitrage capabilities have been tested on 25, 50 and 120 day long positions.

Our results, shown in Figures 7.8 and 7.9, indicate that ISOG leads. In particular, Figure 7.8 displays ISOG arbitrage capabilities on different day long positions. We

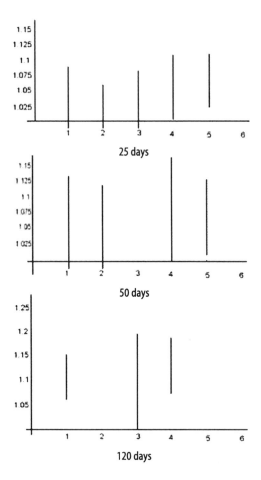

25 days

50 days

120 days

Figure 7.7. Fluctuations in maximum gains in the case of SOM (first three bars) versus ISOG (last three bars).

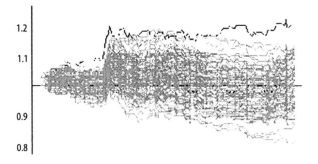

Figure 7.8. Results obtained by operating random decisions in the market (gray lines) are compared with those obtained by joining ISOG suggestions (black line). X-axis shows time, Y-axis shows the payoff under the hypotheses of compound interest law.

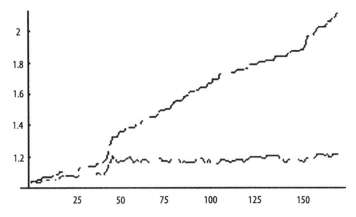

Figure 7.9. A comparison between the results of the net (gray line) and those obtained by insiders. Time is shown on X-axis, payoff on Y-axis.

have assumed c belonging to the range [0.001, 0.005] and i varying in the range [0.08, 0.2]. Filled areas represent the combinations of transaction costs and passive rates that ISOG can hold. The larger the area is, the better is the performance the net has offered. The uncovered region doesn't exceed respectively 12%, 5% and 8% of the whole area, according to whether 25, 50 or 120 day long positions have been considered.

The results prove that ISOG can sustain cost levels up to 0.005 for each transaction, as well as borrowing rates up to 20% annual rate (really near the limit considered usury conditions by the Bank of Italy). These results were compared with those obtained from using random decisions for going in and out of the market. This test helped us to prove the goodness of the results we obtained, and more specifically that the results obtained did not depend exclusively on the strategy but on ISOG prediction capabilities.

Figure 7.9 shows this idea in detail. Gray lines indicate results we got by joining repeatedly random decisions, while the black line represents the payoff of the net. The gap between random decisions and ISOG performances tends to increase over time. The latter test, finally, helped us to verify how much our results differ from the ones of insiders, giving us a sort of measure of profitability for our method. Gains are calculated according to compound interest law. It is easy to see that overall results are more satisfactory for earlier days belonging to a selected temporal frame, when net performances equal the ones obtained by insiders. We think these results could be improved by modifying the ISOG learning parameters (α, *reco* and *threshold*), and more specifically by acting on neuron replication capabilities: the higher they are, the higher the final number of neurons is. This means an addition in model complexity, but also an increase in overall performance, because the probability of uncovered patterns being correctly recognized should increase together with the number of neurons in the map.

Performances are similar in the earlier 50 days, then insiders' results go up more rapidly. The latter test, finally, helped us to verify how much our results differ from the ones of insiders, giving us a sort of measure of profitability for our method. Overall results are more satisfactory for earlier days belonging to a selected temporal frame, when net performances equal the ones obtained by insiders. We think these results could be improved by modifying the ISOG learning parameters (α, *reco* and *threshold*).

7.4 Conclusions

In this chapter we studied the possibility of employing ISOG-based nets as an alternative to traditional models in financial markets forecasting. We presented a review of unsupervised neural network models including the principal aspects of traditional self-organizing maps and introduced the Integrated Self-Organization and Genetic (ISOG) algorithm, which we think is more suitable to deal with financial applications. We showed the ISOG performance both as a classifier and for forecasting financial markets. We referred to a practical case, comparing maps (SOMs and ISOGs) on daily data from the Milan Stock Exchange Official Index. We proved that ISOG-based nets are more stable and produce more satisfactory results than SOMs. ISOGs offer arbitrage opportunities in the market: this net can sustain high operational cost levels and borrowing rates close to the usury line fixed by the Bank of Italy. ISOG appears to provide robust nets: we think the results offer a good starting point for future research involving changes in the values of fundamental ISOG parameters.

8 Real Estate Investment Appraisal of Land Properties using SOM

Eero Carlson

Abstract

Real estate market analysis is a complex process: each property is unique, land use may be controlled, sale prizes are recorded maybe once in 30 years, and finding comparable sales is not easy. Nevertheless, some understanding of the resemblance of the properties is necessary. Self-organizing maps are used in this chapter by E. Carlson to analyze real estate data from the National Land Survey of Finland. Geographic location data is included in the mapping process. The application was developed using object-oriented Smallworld GIS software. The input vectors were collected using external databases on properties and buildings, and digital map databases.

8.1 Introduction

Real estate market analysis is a cognitively highly demanding process. Each property is unique, land use may be controlled, and sale price is recorded maybe once in 30 years. Finding comparable sales is not easy. Nevertheless, some understanding of the resemblance of the properties seems to exist. Some ordered insight can be achieved by observing a sufficient number of properties or real world *objects*. A self-organizing process can then be used to create an understanding of typical objects as well as an understanding of less typical or suitable objects in a particular neighborhood.

The components of the objects vary: location, size of property, age of a building, nearness to sea or lake, size of lake, etc. Future cashflow calculations are important for property units used in business. Varying living situations are important for residential and recreational sites. There is no objective way of selecting the components or of specifying the importance of one individual component. One looks at a property for a specific purpose and the sale price is always determined only by two participants, the buyer and the seller.

A self-organizing map of property units can create an ordered visualization and understanding of a market situation. The processing unit or neurons on a SOM learn and adapt the market and property unit information. The identity of the unit is determined only from the observations matching to each neuron. More important are the topologically neighboring units, the observations most similar to the object that are just a bit different in possibly some aspects.

8.2 Geographic Information Systems

A natural way to manage geographic information is to use a Geographic Information System (GIS). Application development at the National Land Survey of Finland is based on object-oriented Smallworld GIS and Smallworld Magik software[1] (Carlson, 1992).

The only necessary observations for GIS are location and sale price, all other components can be undefined. Scaling constants are given to each component. The similarity between objects is computed using the Euclidean distance between the vectors of the objects and the scaled values of the components should be reasonable.

The SOM algorithm organizes the observations and the resulting neurons are loaded into the data sets of GIS. Each observation is chained to the best matching neuron, thus the neuron can be used to get the observation list.

For data management purposes the scaling variables are used so that specialized views of the observation vectors are created. One map is organized so that the geographic location is dominating. In this case the SOM divides the geographic space into polygons where matching observations are grouped together. The topology of the map is fully ordered according to the dominating components; the other components are averaged.

Accordingly some other components are chosen and the resulting SOM provides a useful interface to search the data sets, giving only values for these components. One should be aware of the scaling process and the purposes of the organizing. The original idea of the SOM is to find the best representation of the observation space. If some components are overstressed then the view tends to be narrower relative to other components.

The most important aspect in the data management is that the SOM preserves the distribution of the observations. The neurons are used as clickable buttons to the observations and each button usually refers to an equal number of observations. If there are ravines on the map then there are neurons with no observations. Similarly, totally equal observations are always matched with the same neuron, but this is not normally possible in real-world observations.

The SOM preserves the topology: if comparable neurons are not found then it is easy to browse over the topological neighborhood and find more or less comparable observations. The distance to the prototype or quantization error can be computed and the dissimilarity can be seen.

A hierarchical solution is proposed to manage very large data sets. For example, in the case of Finland, the whole country with one million property units is first organized as 100 neurons, each containing a group of about 10,000 units, and this group of similar units is then organized to 400 neurons, each with an average 25 matching observations (Figure 8.1).

---

[1] These are trademarks of Smallworldwide Ltd., Cambridge, England.

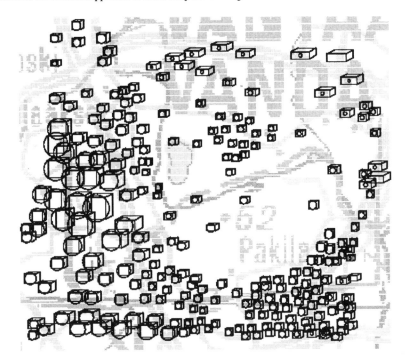

Figure 8.1. Building database as a SOM with 252 neurons representing 8842 buildings in Helsinki, Espoo and Vantaa. Location is dominating. The symbol shows averaged values of buildings. The area and the number of floors are displayed as a box and the number of inhabitants as a circle. Observations provided by the Population Register Centre of Finland. © National Land Survey of Finland.

Another solution is to use very large SOMs. One effective example is the WEBSOM method [see Honkela et al. (1996) and Chapter 12]. Two-level organization of full-text documents creates an ordered document map and original documents are seen with no more than two or three button-pressings. Any subarea of the map can be zoomed and the desired documents are listed by clicking the neuron. Topologically neighboring documents are easily found.

8.3 Visualization

Traditional cartographic knowledge can be used to design the symbols for the neurons. Graphic variables can be used for each component. Location (x, y), color (hue, saturation, intensity), size, shape, texture and orientation can be used for a suitable combination of components. Visually greater values are used as greater values of the component. Location, size, intensity, saturation, texture and orientation represent relationships of order and proportion; hue also represents relationships of order, but in addition it represents strong relationships of kind. Shape represents only relationships of kind. One example is given in Figure 8.2.

The observation vector contains the geographic coordinates. In the self-organization

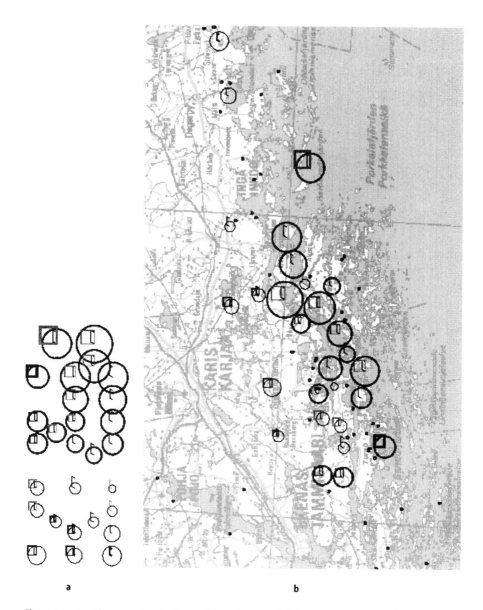

Figure 8.2. Seaside recreational sites in Inkoo. The symbol has three parts and shows 12 of 24 components. First, the area of the circle shows the price up to FIM 1,230,800 (1 US dollar is about 5 FIM) and the thicker line means shoreside. Second, the area of the square shows the floor area of the building (up to 159 m²), the thicker line means more build-ups, and light gray is for older buildings. Third, the area of the parallelogram shows the parcel area, the thicker line means official coastal plan, and light gray means parceling out not done. Full topological order can be seen for shoreside, building, build-ups, coastal plan and parceling out. **a** Neurons with no observations are not displayed. Symbol locations are slightly edited on the geographic map. **b** Observations (53) are shown as small circles. © National Land Survey of Finland.

a   b

process these coordinates are handled similarly with all other components. Each neuron can be visualized on the geographic map at its location. The distribution of the observations is then seen. Objects with many similar observations are represented by many neurons with only slight differences, but exceptional cases are represented only by one or a few neurons in the average location of these observations. This should be known to avoid misinterpretations. The topological neighborhood gives information as to how good the prototype object represented as a neuron is. The deviation of observations of the matching neuron can be compared with the deviation in the nearest topological neighborhood.

To look at the conceptual categories in the observation space ordered display of the SOM neurons is very useful. The categories are learned only from observations and the categories are organized so that similar ones are usually neighbors. Thus, the conceptual hierarchy can be formed by dividing the groups of neurons at an arbitrary level of abstraction. Outliers can be seen clearly and those observations affect only the matching neuron and the neighboring neurons.

8.4 Scaling

Real estate market analysis is not easy. Property units are unique. Participants in the market are unique, too. In many cases it is not possible to find an ideal property. Expectations about the future change all the time and change very rapidly. Some property units have potential for different developments, some have not. Living situations are changing and differing components are critical for residential, recreational and production sites.

The problem of strongly varying situations in the market is solved using scaling of the measured components. Differing views from the same observations are generated by producing SOMs with slightly varying scaling constants of the components. The larger the scaling constants the stronger the resolution or number of neurons for a component. Accordingly, smaller scaling constants give more averaged values on the neurons or a narrower view in the direction of these components.

Good values to begin with are to normalize the minimum and maximum between 0 and 100. In many cases, there are some exceptionally high values, sometimes gross errors. It is not advisable to give power to only one exceptional individual observation.

Other good values to begin with are to normalize the variance of the component values. This measure is sensitive to some exceptionally high values; power is given to a small minority of individual observations.

The organized maps are compared by using the quantization error of the learning set and the deviations on each component can be computed, too. Visualization of the map gives more information of the conceptual groups and possible topological problems. The scaling changes the topological order of the map slightly. Strongly scaled components can be seen as compact regions in topological order, especially in the case where these components are the main components in the data set. Weakly scaled components are then seen forming subregions within the main regions.

A measure of the goodness is proposed in Kaski and Lagus (1996). The goal is to find SOMs where unnecessary dividing of regions and distributing of similar observations to quite different regions is avoided. This is not an easy job. Similar observations are in many cases in "great valleys" where the distances on the map lattice between similar neurons can be large. Accordingly, exceptional observations on the ravines may match the same neurons and the distances between non-similar neurons can be small.

Maybe the most important and useful thing in appraisal is that one object can be seen in more or less differing topological regions on the maps depending on the importance given to chosen components. Contradictory situations can be discovered. Comparable sales can be chosen in many ways and some objective truth can be understood as a combination of two subjective truths.

8.5 Sensitivity Analysis

To understand the goodness of the SOMs, we analyze the sensitivity of the components in relation to changes in other components. This is implemented by computing an array of values for a chosen component. Two other components are chosen for both axes and a suitable window and the number of rows and columns. The value of the third component, usually the price in appraisal cases, is displayed as the color of the pixel and as a numeric value as desired (Carlson, 1997).

Neurons are seen as polygons if one single object and one single map is used.

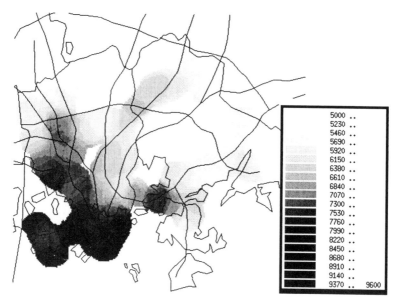

Figure 8.3. Price topography of two-room dwellings in Helsinki. Geographic space is used to analyze sensitivity in price formation. The data consisted of a portfolio of five dwellings from 40 to 60 square meters floor area. Four slightly varying SOMs are used. Average price per square meter is computed for 240 rows and columns. Only pixels within a tolerance of 1.6 compared with the average quantization error of the learning set are displayed. Values are not shown in regions for one-family houses and row houses, as seen in Figure 8.1. Observations from January 1992 to March 1995 provided by the Technical Research Centre of Finland. © National Land Survey of Finland.

Corresponding neurons can be highlighted on the SOM. Many neurons are seen in the direction of strongly scaled components and only one or a few in the direction of weakly scaled components (Figure 8.3; Table 8.1).

The quantization error is computed and the display of the value is controlled by accepting only those pixels where the quantization error is less than a proportion of the quantization error of the learning set. Strongly scaled components tolerate only small changes in their values and weakly scaled components are accepted with quite wide variation. Only those regions are displayed where sufficient numbers of resembling observations are seen.

Especially interesting neurons give discontinuities on the display. These neurons with their matching observations are compared and in many cases some more important components can be uncovered. In the real-world phenomena, one component does not have an exact global meaning. The meaning emerges in a given situation, in a given combination of component values and for a given participant on the market. The effects can be generalized by organizing similar observations.

To understand the difference between neighboring observations, recall that the SOM algorithm produces neurons where the observations in the neighborhood are essential. The distribution of the values of a given component is dependent on the scaling constant. The averaged values for this component are closer to each other on weakly scaled components and the boundary area is quite large.

The boundary effect and the role of scaling can be seen clearly by organizing observations on a linear grayscale. The result should be average gray. By scaling this component strongly, all values between black and white can be seen but the SOM can be very confusing. The effect of principal components cannot be destroyed and minor components are averaged more as necessary. By weaker scaling, the boundaries are growing and the values are more or less gray and the number of neurons is decreasing. Weak scaling gives only one or two neurons as average gray.

Simultaneous use of a group of maps with a varying scaling constant is one possibility to smooth the representation of the function to be visualized. The number of neurons used in a given window of component values and the points where the neurons are changing are varying. The averaged value from a group of maps gives a smoother view of one particular component. Suitable maps used can be selected in the application.

8.6 Portfolio Computation

Another way to smooth the quantized values given by SOM is to use a group of objects accessed as a portfolio. The best matching neurons give values for each object and these values are summed and divided by the number of objects in the portfolio. Slightly different objects take values from neurons in the neighborhood or in other subareas and maybe better values are computed for objects between typical cases.

Portfolio computation is especially useful in sensitivity analysis. For each object, for each pixel, with each map, the value of one component is computed and the resulting average value is displayed as color or numeric. The typical combinations of component values belonging usually together are seen.

One important use of a portfolio is in analyzing rare components and exceptional cases. Some situations are present only occasionally and the distribution of observations is so rich that no clustering is possible within the observation space. These objects can be compared with the typical values without the exceptional situation present. Using suitable grouping of these objects, the values of the portfolio can be compared and the average effect can be shown.

The goodness of the SOMs can be compared by computing the portfolio values for given subtypes in the observation space. Some maps may be dedicated to special types or to special components and the results should be better in those special cases. Some systematic errors are sometimes seen, especially on the boundaries of the observation space, where the essentially important neighbors are missing on the other side.

8.7 Adaptation to New Observations

There are some interesting issues: how to manage increasing information over time; how to understand the structural changes in the market; how to adapt to changing prices; how to see the up- and down-signals. Information from past situations is needed. Price indexes are important. The groups or subtypes of the real-world objects should be relevant for user needs.

One possibility is to use sales date as a component of the object. The SOM is organized and topological regions or subregions can be seen. In many cases this component is one of the most important components. This can be seen in the ordered display. This kind of map cannot adapt to new observations, which are allocating quite new areas of the observation space. The new SOM is newly created and the structures may differ in the topological order of other components.

Another possibility is to use some fixed time window and to organize the map without using sales date as a component. One example was shown with an eight-year series of shore parcels (Carlson, 1991). The conceptually organized SOM adapts new observations using a moving time window and both structural changes and price changes can be seen easily. Only fine-tuning of the map is needed in the self-organization process and the main topological order does not change. Price indexes are computed using a given portfolio and the price of the portfolio is computed using successive maps with successive time windows.

Market changes can be detected sensitively. New observations can be compared with the values given by the SOMs. One object is always compared with the most similar one. In this way the method gives up- and down-signals sensitively. This is similar to the moving average method used for stock market changes.

8.8 Other Examples

Forest woodlots were analyzed in Carlson (1990). In this analysis of woodlots the location, timber volumes, the structure of the woodlot, and the distances to a road and to a town are represented as 14 components. The principal topological order is determined mainly from the dominating age of the growing stands: clearcut areas with full age stands and growing timber volumes have increasing prices; possibilities to cut are very limited for young stands and negative cashflow reduces the price.

One early experiment involved the distribution of woodlots into two classes. In this the forestry components were scaled properly. Average results for administrative regions were produced by scaling principal components down and by giving great importance to administrative codes. The narrow view of the average values of administrative regions can be demonstrated clearly. Proper use of the SOM gives, however, a richer view and the diversity of observations can be seen.

Remote sensing data classified from satellite images is useful to analyze the woodlots. Land use classes for 25 meter pixels were collected for each woodlot and 64 components were added to the observation vector. These additional components do not change the main topological order of the SOM. One additional region can be seen. Woodlots with large timber volumes are divided into two parts and satellite images taken later indicate their clearcut areas.

Shore parcels without buildings were analyzed in Carlson (1991). In this analysis the following components were used: geographic location, size of the lake, distance of sight, site on mainland or island with/without road or river, direction to lake, official coastal plan, parcel area and the sale price. Principal topological order is determined by size of the lake, distance of sight and official coastal plan. The effect of the distance to Helsinki is clearly seen on the geographic map.

Eight years (1982–1989) of shore parcel observations were divided into eight consecutive time windows. The year 1987 (with most observations) was chosen as the beginner of the learning. The organized SOM was then fine-tuned with each consecutive window from the year 1982 to 1989. The new SOM for the year 1987 had some slight topological differences. The official coastal plan was in one region and island observations were divided into two separate regions.

Structural changes in the market can be seen on the geographic maps and on the SOMs. Price changes can also be seen clearly. Observations were compared with the prices from the SOM and a down-signal was detected in the year 1989. Each observation is compared with the best matching neuron and that is why SOM gives the change information in the market sensitively.

Dwellings in Helsinki were analyzed in Tulkki (1996) in co-operation with the Technical Research Centre of Finland. The main components used in this analysis were: geographic location, price, number of rooms, floor area, year of construction, condition, number of floors in the building, floor, elevator, and nine other components were used. Many components have not much to do with price. For example, the number of floors can be seen clearly on the SOM but the prices are determined mainly by other components.

Special attention was paid to analyzing the geographic location (Figure 8.3). A portfolio of two-room dwellings was computed over the whole area of Helsinki and the price topography can be seen. The highest prices can be seen in the center where the buildings are the oldest. Accordingly, the lowest prices can be seen in the newest regions. Slightly lower prices can be seen in eastern parts compared with western parts. Railways to the north-east and to the north-west are clearly seen as higher prices, too.

The observations were from 1992 to March 1995 and the market changes were moderately strong. The price component was scaled moderately weak to get average prices over the whole period. Another reason to use weak scaling for price is to find the price for a new object using SOM with other components. The observations were divided into time windows for the years 1992, 1993 and 1994. A sensitivity analysis shows price changes in two-room dwellings in the center of Helsinki (Table 8.1). The nonlinear form of the function can be clearly seen.

Recreational parcels were analyzed in Lehtonen (1996). Special attention was paid in this analysis to buildings and shores. This analysis used building data from the Population Register Centre of Finland. The size, age and build-ups were represented by 11 components. Geographic location, price, parcel area, information of official coastal plan and parceling out were important components. Shore data was computed using digital map data at scale 1:250,000. For each parcel, the distance to the nearest water area, the size of the lake and the direction to the lake was computed. Distance to electric power line area was also computed.

Price topography was computed using six slightly varying SOMs for groups of objects. The lowest prices are seen for parcels without shore, without building and where parceling out is not done. The highest prices were found in the Inkoo region (Figure 8.2), where seaside and large buildings with amenities were the most desired objects. The observations were rare but the advantage of using SOM could clearly be seen from this example. Organized views of 53 observations, from 53 exceptional cases, are displayed as 50 neurons and the importance of the topological neighbors can be seen. The most similar cases are nicely related to each other and some understanding of the dependencies can be created. If we use 10 to 20 observations per neuron to get good statistically averaged values we need some 500 to 1000 km shore and the geographic location becomes dominant in price formation. This type of map shows the highest prices also in the Inkoo region, but the prices are much lower.

Table 8.1. The price (FIM per square meters) of two-room dwellings in the center of Helsinki in the years 1992, 1993 and 1994. The year of construction is on the horizontal axis and the condition of the dwelling is on the vertical axis.

1890	1900	1910	1920	1930	1940	1950	1960	1970	1980	1990	Condition	Year
9041	9166	9179	8875	9014	8991	8639	8618	8785	9155	9326	good	
8446	8383	8482	8228	8049	8128	8066	8088	8133	8505	8630	normal	1992
..	8315	7950	7652	7562	7556	..	..	..	..	..	poor	
9858	10253	10335	9661	9171	8948	8929	9194	9364	9695	9731	good	
9359	9469	9352	8310	8256	8036	7937	8270	8193	8427	9078	normal	1993
8683	8545	8294	8019	7821	7610	7639	8361	8303	..	..	poor	
10900	11280	11347	10463	10427	10208	10004	10538	10679	10620	..	good	
9661	9774	9779	9595	9429	9103	9531	9955	10213	10442	..	normal	1994
..	10275	10214	9657	9250	9162	..	..	..	..	..	bad	

Year of construction (spanning columns 1890–1990)

Distance to electric power line was analyzed using portfolio computation (Carlson, 1997). The distribution of the rare observations is so that no good clustering is possible with SOM. These observations were collected as portfolios grouped according to the distance and to shoreside and to building. These portfolios were computed using the SOMs and were compared with the values of the portfolios. The effect was FIM 10,000 to 30,000.

The sensitivity of parcel areas was analyzed (Carlson, 1997) using six SOMs. Two maps represent increasing prices for increasing parcel areas as expected. One SOM represents decreasing prices for increasing parcel areas. Other maps show no dependency. This component is contradictory and both parts can be demonstrated using the scaling of components. It seems that the most demanded land is divided into smaller parcels and the less demanded land is offered in larger parcels.

8.9 Conclusion

Real estate market analysis is a complex business. The SOM approach is very useful for comparison and appraisal of properties. Finding suitable properties or property attributes (components) is the most critical point in understanding market behavior. Plenty of data is available in many public databases and a vast amount of experience and knowledge can be used. Nevertheless, market behavior can be understood only by organizing observations from a real-world perspective. The observation space is sparse, with many nonlinearities and discontinuities. Nevertheless, it is important to understand the situations and components belonging to each other and to discriminate between a huge number of special cases. This chapter shows how the diversity of components can be scaled for differing purposes. Contradictory components and objects can be seen from the SOM maps.

Furthermore, market changes can be imbedded using moving time windows and adapting organized SOMs to the structure and price level of new observations. Price indexes can be computed using a given portfolio and the series of SOMs. Up- and down-signals can be detected by comparing the newest observations to the values given by the SOM.

9 Real Estate Investment Appraisal of Buildings using SOM

Anna Tulkki

Abstract

The most commonly used methods to analyze and value real estates are econometric models. However, these models have some weaknesses that make it difficult to obtain good analyses or reliable models. First, there are the linearity assumptions, and second the problem of correlating variables. Furthermore, changes in real estate and dwelling prices in the 1990s have made it especially difficult to use econometric models. The aim of the research presented in this chapter is to show how the self-organizing map is more suitable for appraisal of the prices of dwellings and/or for wider use in real estate valuation. The reason is that the self-organizing map is a neural network technique that allows to examine non-linearity as well as providing a capability to preserve the topology and distribution of the data, which is very important in real estate valuation.[1] The application in this chapter involves the Finnish dwelling market.

9.1 Characteristic Features of the Finnish Real Estate Market

Some typical features of the Finnish dwelling markets need to be known in order to be able to follow the analysis that follows.

In Finland about 33% of the people live in apartment houses.[2] Of all dwellings only one-third are subsidized and the rest are privately financed. It is very common that the residents own the dwellings that they live in. After the war owner occupation became popular and it still has a very strong base. However, the deep depression that took place at the beginning of the 1990s has increased the demand for rental dwellings, although it seems that many young people want to get a dwelling of their own.

The owners of the dwellings in an apartment house form a condominium or housing association. The purpose of the condominium association is to maintain

---

[1] This article is based on the author's Master's thesis at the Department of Surveying of the Helsinki University of Technology. This research was a cooperative work with the Technical Research Centre of Finland and the National Land Survey of Finland.

[2] Statistic value 12.3.1995, Central Statistical Office of Finland.

the apartment house. To do this a maintenance charge is collected from each owner each month. The amount of the charge is determined by the condominium and it often depends on the condition of the house: if there is repair work in prospect or it has just been done, the charge can be more than it usually is. The amount charged is determined by the number of square meters, so owners who have large dwellings pay more than those who own small ones. The average maintenance charge is about 12 FIM (or $US 2.5) per square meter per month.

The condominium association can loan money to finance repair works as well as building costs. The condominium's debt is paid back by the maintenance charge or it can also amass and be paid back in portion to the debts that the dwelling owners are supposed to pay. If a dwelling is sold before the debt portion is paid back the new owner is responsible for paying the debt.

Apartment houses can either lie on land that is rented from the city or the condominium can own the land. The rental deal is usually something like 60 years.

The prices of the dwellings are in no way restricted. The owner can ask any price and a purchaser can offer the price he thinks is suitable. So the final price is based on their mutual agreement. The sales we studied were all, however, made with real estate brokers and this may have some effect on the final price. The range of prices of dwellings is usually rather wide.

9.2 The Data

The Technical Research Centre of Finland (VTT) has collected prices and features of dwellings since 1977. The data is provided by the largest real estate agents in Finland. For this study we chose the purchases of second-hand dwellings of apartment houses that had taken place in Helsinki during the period from January 1, 1992 to March 31, 1995.

All the variables were taken from the original data with the exception of a few that involved the lot and included mostly missing values. The total number of variables was 19.

After some elimination had been done the data included 11,301 purchases. Characteristic parameters of the variables that were examined are presented in Table 9.1.

The data was divided into two parts: training and testing. The training data included 90% and the test set 10% of the purchases. It rapidly became evident that it was impossible to make a neural network with 1000 neurons that would represent the whole data set so that appraisals could be made from one model.

The next step was to divide the initial data set into smaller parts so that they in turn were more suitable for self-organizing mapping. We chose to segment the data set so that subsets were all the sales of one-room dwellings, all the sales of two-room dwellings, and all sales during the time period January 1, 1994 to March 31, 1995.

9.3 Preprocessing of the Data and the Research Method

Some purchases in the data included missing variables (e.g. *age of the building, lot, vacant or rented, existence of an elevator, condition, floor, number of floors,*

Table 9.1. Statistics of the variables

Variable	Samples	Missing samples	Minimum	Maximum	Range	Average	Variance
Age of the building	10706	595	1	221	220	42.29	22.875
Lot (1=not rented, 2=rented)	9579	1722	1	2	1	1.25	0.433
Number of rooms	11301	0	1	5	4	2.06	1.022
Area (m²)	11301	0	12	296	284	54.43	28.047
Vacant=1, Rented=2	9439	1862	1	2	1	1.06	0.240
Existence of an elevator (1=exists, 2=none)	9531	1770	1	2	1	1.5	0.5
Condition of the dwelling (1=good, 2=acceptable, 3=poor)	10601	700	1	3	2	1.82	0.616
Floor	10548	753	1	9	8	3.01	1.654
Number of floors	9307	1994	2	9	7	4.85	1.723
Maintenance charge (FIM/m² per month)	11292	9	0	90	90	11.85	5.642
Price (1000 FIM)	11301	0	70	4000	3930	394.02	259.849
Debt (1000 FIM)	11301	0	0	609	609	3.84	16.054
North-coordinate (m)	10818	483	6672534	6687478	14944	6677978	3241.43
East-coordinate (m)	10818	483	3380492	3397514	17022	3386575	3633.50
Square price (FIM/m²)	11301	0	2932	23000	20068	7385.30	1895.390
Number of selling days	11252	49	0	490	490	47.83	52.047
Date of selling [a]	11301	0	11689	12873	1184	12293.00	326.801
First date of selling [a]	11252	49	11338	12871	1533	12245.19	333.333
Distance from downtown (m)	10818	483	192.00	13556.43	13364.43	4588.66	3474.38

[a] Values are integers from SAS-program.

north- and *east-coordinates, distance from downtown, number of selling days* and *first date of selling*). These variables were input as missing components, which were indicated by the letter "X" for the neural network program. It has been found that the results are often better if the samples that include missing values are used than if they are left out.

The initial variables differed greatly from each other: for example, the existence of an elevator could have values 1 or 2 but the price varied from 70 to 4000 FIM. Variables that have wide ranges are stronger in the organizing process than those with small ranges. For that reason the variables had to be scaled before training the map.

The best map that would estimate the prices of dwellings was searched by trying different kinds of scaling of the variables and also by trying different training parameters. The results of the training process were checked by the quantization error method and Sammon mapping. Then the better organized maps were tested by the test data, without the knowledge of the square footages and the total prices. The criteria of goodness of the maps were the correlation between the appraisals of the dwelling prices and the real ones. Also, the average deviation was calculated.[3]

---

[3] All computations were made with the SOM_PAK program (Kohonen et al., 1995).

9.4 The Results

As stated before, the data was divided into smaller subsets because of the difficulties of representing it on one map that would be appropriate for appraisal. The best map, i.e. the best correlation between the prices of real sales and the appraisals, was achieved by the map that represented the sales during the time period January 1, 1994 to March 31, 1995. The reason was that big changes in prices occurred during the research time period 1992–1995, as is evident from Figure 9.1.

Besides time, the most significant factors in all the data sets were the location (*north-* and *east-coordinate* and *distance from downtown*) and the size of the dwelling (*number of rooms* and *area*). In Helsinki, location is notoriously a very important regressor; even in a small area in a neighborhood, prices can vary a lot because of the location. In addition many factors correlate with the location variables.

The correlation of the total prices of the map from January 1, 1994 to March 31, 1995 was 0.93 and of the square prices 0.80. The average deviations were 854 FIM/m^2 and 57,000 FIM. The averages of the real prices were 7854 FIM/m^2 and 399,000 FIM. The Sammon projection of this map is presented in Figure 9.2.

The correlation of the square prices of one-room dwellings was 0.67 and of the total prices 0.85. The deviations were 1022 FIM/m^2 and 32,000 FIM. The averages of the real prices were 8051 FIM/m^2 and 242,000 FIM.

The correlation of two-room dwellings square prices was 0.74 and the deviation 826 FIM/m^2, while the average square price of two-room dwellings was 7042 FIM/m^2. The results of total prices were 0.81 and 46,000 FIM, the average of total prices was 362,000 FIM.

One map was made of all the sales that represented the average prices for the time period 1.1.1992 to 3.31.1995. The time variables and the price variables were scaled so that their weights were much weaker than the rest of the variables. The variances of the variables are shown in Table 9.2.

The map consisted of 36 by 28 units on a two-dimensional hexagonal lattice.[4] A picture of the Sammon projection is presented in Figure 9.3. The correlation of the prices was 0.90 and of the square prices 0.70. The average deviations were 993 FIM/m^2 and 63,000 FIM. The averages of the real prices were 7379 FIM/m^2 and 387,000 FIM.

9.5 Component Planes of the Map

In the component planes, Figure 9.4a–e, white denotes the largest value; black the smallest value and intermediate values have been encoded by shades of gray.

The variables that had the strongest effect on the organizing process have organized on the diameters. The variable *number of rooms* and the variables that

---

[4] The form of neighborhood function was bubble and the learning-rate factor was inverse-time type. In the first phase, the neighborhood radius diminished from 34 to 1 in 55,000 steps. In the second phase, the neighborhood radius diminished from 4 to 1 in 550,000 steps. The coefficient a(*t*) decreased from 0.2 to 0 in the first phase and from 0.04 to 0 in the second phase.

Table 9.2. Variances applied to the variables
for the single map of all sales

Variable	Variance
Number of rooms	1.00
Area	1.00
Price	0.04
Age of the building	0.20
Maintenance charge	0.10
Debt	0.06
Number of selling days	0.10
Distance from downtown	1.00
North-coordinate	1.00
East-coordinate	1.10
Square price	0.04
Vacant/Rented	0.10
Lot	0.10
Existence of an elevator	0.10
Number of floors	0.10
Floor	0.10
Condition	0.10
First selling date	0.03
Date of selling	0.03

correlate with it, *area* and *price* have organized on one cross-stratification diameter and on the other is *distance from downtown*. The correlation between *east-coordinate* and *square price* is obvious: in the east square prices are lower than in the west, a fact that is generally known. The component planes *number of floors* and *floor* are very much alike and the variable *existence of an elevator* makes a mirror image of them. Because all of these variables have had small weights in the organizing process, they are all much averaged. The variable *floor* does not enhance information about the dwelling prices and so it might be better to examine if the dwelling is located on the lowest, the highest or somewhere in the middle of the floors.

The component planes *lot* and *distance from downtown* provide evidence that buildings that locate in the downtown or near it are more likely to lie / stand on land that the condominium owns than those which locate far from the downtown. Also, it is discernible that the buildings in the downtown are older than outside of it, see planes *age of the building* and *distance from downtown*.

The planes *first selling date* and *date of selling* are quite similar and the difference between them is shown on the plane of *number of selling days*. Note that these variables had very small weight in the organizing process. This is also discernible from the values of dates, for example, *date of selling* ranges only from 5.31.1993 to 2.12.1994.

Comparing the planes *debt* and *number of selling days*, it can be observed that the duration of selling is often longer if a large debt is due to be paid for the dwelling. *Maintenance charge* is independent from the other features. However,

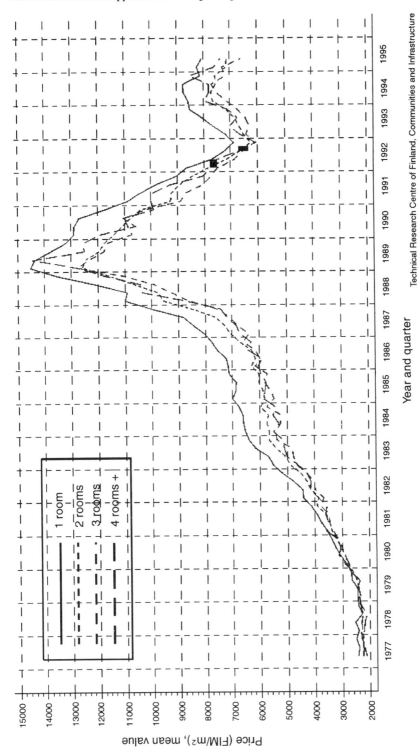

Figure 9.1. Development of second-hand dwelling prices by dwelling type in Helsinki 1977–1995.

Technical Research Centre of Finland, Communities and Infrastructure

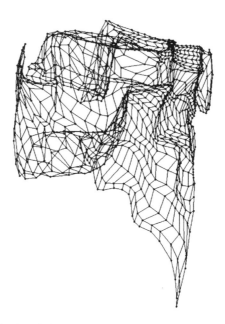

Figure 9.2. Sammon projection of the map describing dwelling sales during the time period 1.1.1994 to 5.31.1995. Dots denote projections of the model vectors on a two-dimensional plane, and lines connect projections of the model vectors of neighboring map units.

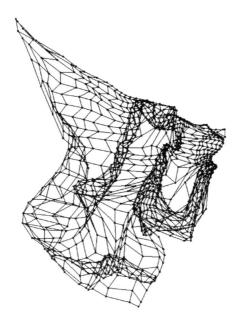

Figure 9.3. Sammon projection of the map describing dwelling sales during the time period 1.1.1992 to 5.31.1995.

if the debt is large sometimes the maintenance charge is also high and so also duration of selling can be long.

The average deviation of the square prices of the test set and the values that were achieved by the neural network was about 12%. The deviation of the total prices was 14%. However, if the debt is high, sometimes the maintenance of the property is also high, and in consequence the time it takes to sell the property can be long.

9.6 Conclusions

The self-organizing map proved to be very suitable for the analysis and appraisal of dwellings. We found that scaling is a very essential part of pre-processing. The result of the training depends on how well the relationships between the variables have been discovered, in other words on scaling.

In real estate valuation, where the aim is to define the probable value objectively, scaling may seem unscientific and dubious. However, the success of scaling can be viewed by the Sammon mapping and the test set. It is feasible to examine different aspects of assessment by changing the scaling factors a little. This feature of self-organizing map should be considered as an advantage; there are no absolute values of real estates, only aspects of the probable value.

A great advantage can further be achieved if the neural network can be linked to a geographical information system (GIS) as discussed in the previous chapter. Thus, the appraiser can easily see the set of similar objects and also locate them on a geographical map. The quantization error can be used as an indicator of the similarity. If necessary, the appraiser could also search more neurons that are close to the input vector, the dwelling of target, or sight the neurons that are geographically near the "winner neuron".

Self-organizing map is very advantageous in analyzing and appraising real estates and so it is worthwhile to aim at wide use in real estate business.

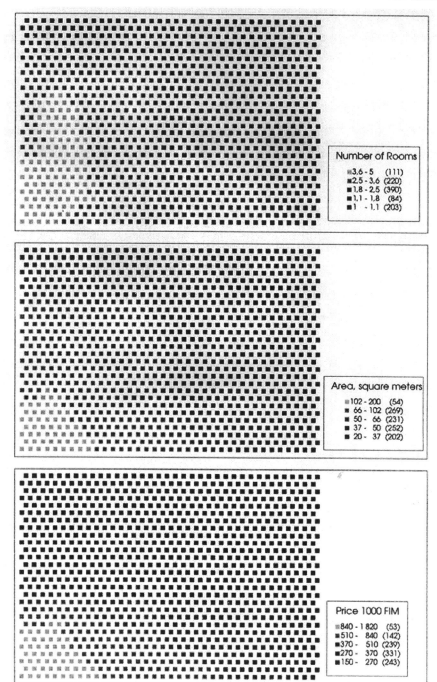

a

Figure 9.4 a to **e**: Component plane of variables influencing property price.

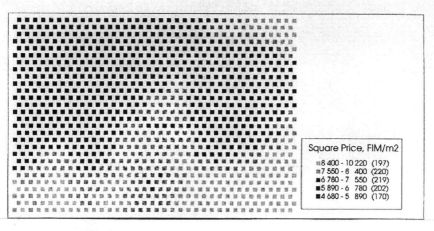

Square Price, FIM/m2

▨ 8 400 - 10 220 (197)
▩ 7 550 - 8 400 (220)
■ 6 780 - 7 550 (219)
■ 5 890 - 6 780 (202)
■ 4 680 - 5 890 (170)

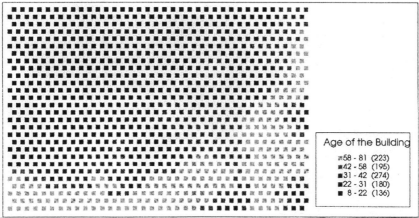

Age of the Building

▨ 58 - 81 (223)
▩ 42 - 58 (195)
■ 31 - 42 (274)
■ 22 - 31 (180)
■ 8 - 22 (136)

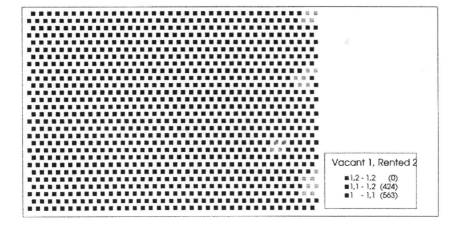

Vacant 1, Rented 2

■ 1,2 - 1,2   (0)
■ 1,1 - 1,2 (424)
■ 1 - 1,1 (563)

b

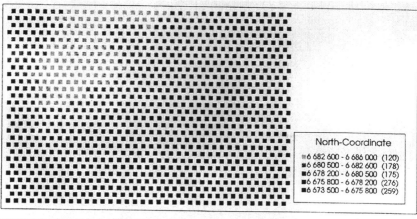

North-Coordinate

▓6 682 600 - 6 686 000 (120)
▓6 680 500 - 6 682 600 (178)
▪6 678 200 - 6 680 500 (175)
▪6 675 800 - 6 678 200 (276)
▪6 673 500 - 6 675 800 (259)

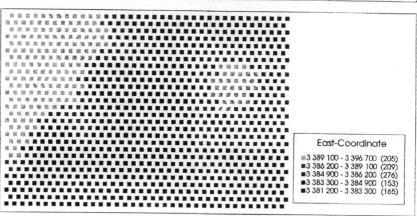

East-Coordinate

▓3 389 100 - 3 396 700 (205)
▓3 386 200 - 3 389 100 (209)
▪3 384 900 - 3 386 200 (276)
▪3 383 300 - 3 384 900 (153)
▪3 381 200 - 3 383 300 (165)

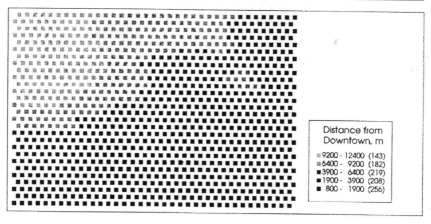

Distance from
Downtown, m

▓9200 - 12400 (143)
▓6400 - 9200 (182)
▪3900 - 6400 (219)
▪1900 - 3900 (208)
▪ 800 - 1900 (256)

c

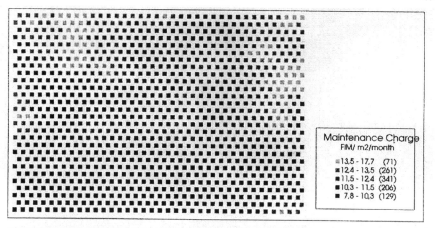

Maintenance Charge
FIM/ m2/month

- 13,5 - 17,7   (71)
- 12,4 - 13,5 (261)
- 11,5 - 12,4 (341)
- 10,3 - 11,5 (206)
- 7,8 - 10,3 (129)

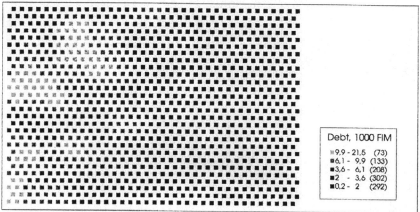

Debt, 1000 FIM

- 9,9 - 21,5   (73)
- 6,1 - 9,9 (133)
- 3,6 - 6,1 (208)
- 2   - 3,6 (302)
- 0,2 - 2 (292)

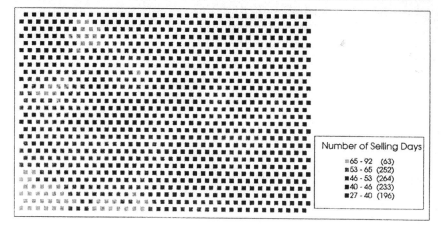

Number of Selling Days

- 65 - 92   (63)
- 53 - 65 (252)
- 46 - 53 (264)
- 40 - 46 (233)
- 27 - 40 (196)

d

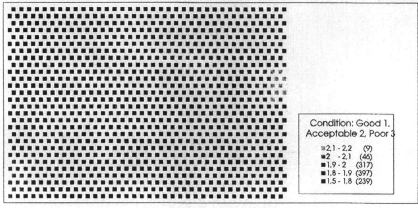

Condition: Good 1,
Acceptable 2, Poor 3

2,1 - 2,2   (9)
2   - 2,1   (46)
1,9 - 2   (317)
1,8 - 1,9 (397)
1,5 - 1,8 (239)

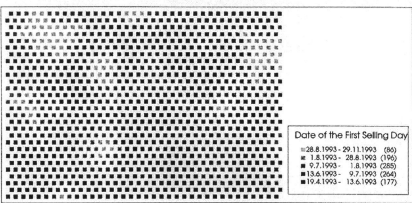

Date of the First Selling Day

28.8.1993 - 29.11.1993 (86)
1.8.1993 - 28.8.1993 (196)
9.7.1993 - 1.8.1993 (285)
13.6.1993 - 9.7.1993 (264)
19.4.1993 - 13.6.1993 (177)

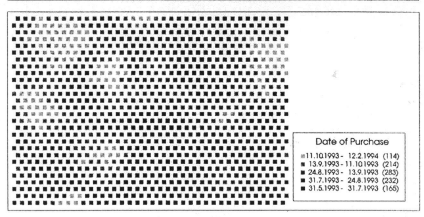

Date of Purchase

11.10.1993 - 12.2.1994 (114)
13.9.1993 - 11.10.1993 (214)
24.8.1993 - 13.9.1993 (283)
31.7.1993 - 24.8.1993 (232)
31.5.1993 - 31.7.1993 (165)

e

10 Differential Patterns in Consumer Purchase Preferences using Self-Organizing Maps: A Case Study of China

Bernd Schmitt and Guido Deboeck

Abstract

Market research and consumer segmentation is still in its infancy in the People's Republic of China. Yet, marketers desperately need information about differential response patterns and consumer segments to formulate viable strategies for the drastically growing Chinese consumer market. This chapter presents the results of consumer surveys with hundreds of respondents conducted in Beijing and Shanghai in the spring of 1997. The data were analyzed using self-organizing maps, a technique that offers easy and convenient visualization of the data obtained without imposing stringent linear constraints on the data. Results indicated important differences in the number and type of consumer subsegments in the Beijing and Shanghai markets with respect to influences on purchase decisions (and, to some degree, lifestyles) but large similarities in media behavior, importance of product attributes, and brand attitudes. The study illustrates the usefulness of the methodology of self-organizing maps. Marketers may use the results of the study as a model of how to collect and analyze consumer surveys to formulate strategies and tactics for the Chinese consumer market.

10.1 Introduction

The current Chinese Consumer Revolution represents a transformation of unprecedented historical proportions, and offers Western companies an unprecedented opportunity to capitalize on events. In order to do so, marketers of consumer goods need to have a strong understanding of China's complex and rapidly-changing consumer market. Until now, our understanding of this market has been limited by the lack of reliable data and methods for analyzing perception and consumption patterns in the marketplace. This chapter presents the data of two large-scale consumer surveys that were conducted in Beijing and Shanghai in the spring of 1997. We analyze the data using self-organizing maps, a technique that offers easy and convenient visualization of the data obtained without imposing stringent linear constraints on the data.

10.2 What do we Know about Chinese Consumers?

Before 1979, when Deng Xiao Ping declared his open-door policy, China's consumers had little choice in the marketplace. They had to consume products of dismal quality produced by state-owned enterprises in a regulated market. Today, China's consumers have a vast selection in a large variety of categories from low-end/budget items to luxury goods. They have access to new media and information. They are exposed to branded goods and their logos and advertisements. New modern retail stores are rising everywhere. So, how do consumers make purchase decisions in this new environment? How often do they watch the new media? How important are different product attributes (e.g., product features, price, quality, the brand image, and the retail environment) to them? What are their attitudes toward brands? Most importantly, are there meaningful differences between different groups of consumers that may be treated as market segments with different wants and needs?

Unfortunately, compared with most other markets, little is known about today's consumers in the People's Republic of China. A major reason for this lack of knowledge is the lack of an infrastructure for market research. Although market research is increasingly being conducted to provide information and intelligence for market decisions, the quality of most research remains questionable. Samples are frequently biased (e.g., mail and telephone surveys are biased toward those consumers that can easily receive mail and have phones); responses can be unreliable (due to untrained interviewers who in the worst case fake data); and consumers have little experience with Likert scales and semantic differentials and show strong inclinations to give socially desirable responses (e.g., in focus-group research). Moreover, most of these data consist of focus group or surveys that focus on broad-scale geographic and demographic segments. Finally, results are often presented in a poor manner and without consideration of patterns in the data and differences among consumer groups. Being able to provide differential responses and relating them to relevant descriptors – in short, consumer segmentation – is of critical importance.

In this chapter, we first review the few segmentation studies that have been published or made publicly available about the Chinese consumer market. Most of this research has been conducted by commercial research firms rather than academic researchers. We then present the methodology and results of a survey conducted in Beijing and Shanghai in the spring of 1997. In contrast to prior research, the data of the survey describe consumer segments on a more finely tuned scale, focusing on demographic characteristics as well as attitudes and lifestyles. Using the self-organizing map methodology, we identify consumer segments for two key markets in China – the Beijing and Shanghai markets. Aside from providing substantive results that are beneficial for marketers who intend to tackle the Chinese market, we show the usefulness of self-organizing maps for exploring survey data based on consumer attitudes and cognitive responses to marketing variables.

The survey is part of a broader research project that focuses on the behavior of today's Chinese consumers in terms of their perception and response to products, brand campaigns, advertising, retailing and other consumption-relevant behavior. This project is conducted at the China-Europe International Business School (CEIBS) in Shanghai. We will therefore refer to it throughout this article as the CEIBS Survey.

10.3 A Selective Review of the Prior Segmentation Research

Most segmentation research done in the People's Republic of China today focuses on geographic segmentation, i.e., the identification of differences between rural and urban consumers, consumers in the coastal provinces and inner provinces, and between the North and the South. This type of research was pioneered by Gallup China in 1994 with a sample of 3400 people, a roughly representative sample of the adult Chinese population. Respondents were interviewed in their homes. The survey revealed sharp differences between rural and urban consumers. For example, results indicated that the urban population was significantly more likely than the rural population to study advertisements before the purchase of a "da jian" – a big ticket item or durable. Moreover, 52% of urban respondents (vs. 38% of all respondents) would pay higher prices for products of high quality. 41% of urban consumers (vs. 30% of all respondents) stated that they would buy a leading brand regardless of price. Brand-name recognition of foreign brands was also the highest in the cities. For example, overall, Coca Cola had 62% recognition, but 94% in the selected cities. The corresponding numbers for Pepsi Cola were 42% for the rural sample and 85% for the selected cities. Finally, the Chinese spent one-third of their income on food; the figure rises to 37% among urbanites and even to 41% in Shanghai.

Dynamics Decision, a research firm, has investigated broad differences among cities. Each month, a sample of 9000 families is surveyed in cities across China on their income level and other demographics as well as consumer-goods expenditure. The average per capita income as well as disposable income is roughly the same for Beijing, Tianjin and Shanghai but more than twice as much for Shenzhen.

Coopers and Lybrand has conducted focus groups in Shanghai and distinguishes three segments based on age and sex: women aged 30–45 appreciate "value and convenience." Men aged 30–45 are "utility shoppers," and consumers aged 30 and under are "highly aspirational and interested in ownership and leisure." Moreover, the focus groups indicate that young women among young consumers are the least concerned about price. Many of them, even if their income is low, spend all of their income on cosmetics and fashion. They favor foreign-invested department stores and boutiques for their atmosphere and service.

Louis Harris, another major research firm, conducted a survey of 2500 consumers in Beijing, Shanghai, Guangzhou, Tianjin and Chengdu. The methodology was based on studies by Yankelovich and Shelly, who are considered pioneers in the measurement of social attitudes in the US. The survey indicated that young, more-affluent and better-educated consumers were much more likely to try new products than older, less-affluent and less-educated consumers. Based on their attitudes and behavior toward new product categories or new brands within a category, they were categorized as the innovators. The survey also found that entrepreneurs and others working in the private sector are more likely to experiment with new products than government employees. Moreover, the young, affluent and educated are more likely to try products admired by others. They have a strong desire to conform to the norms of the reference group.

10.4 The CEIBS Survey

The purpose of the survey conducted at CEIBS was threefold: (1) to provide a more comprehensive survey of the Chinese consumer market in terms of response variables than prior research; (2) to use a reliable methodology and large samples in two key cities, Beijing and Shanghai; and (3) to identify differences between Beijing and Shanghai consumers as well as subsegments within each market.

The surveys were conducted in the spring of 1997 in Beijing and Shanghai by research assistants trained in survey methodology. Participants were given a small present for participation and recruited based on a stratified-sample methodology based on age, gender and income with 30 respondents per cell. Great efforts were made to produce reliable results. To this end, respondents were recruited randomly (following the stratification scheme) but had to complete the survey in rooms that were rented in downtown Beijing and Shanghai in the presence of the research assistants. This procedure provided a safeguard against the practice of faking data by either respondents or field assistants who may have asked somebody else to complete the survey or complete the surveys themselves.

The survey took about 45 minutes to complete. Respondents were asked a variety of questions, mostly on seven-point rating scales, regarding

- attitudes toward advertisements and media;
- influences on purchase decisions;
- media behavior;
- brand attitudes;
- brand loyalty and perceptions of brand parity;
- attitudes toward products of different firms (e.g., state-owned enterprises, joint ventures, imported goods);
- shopping behavior;
- consumer lifestyles.

Moreover, a variety of demographics were assessed, such as sex, education, type of occupation, sector of occupation and income. Not all responses will be discussed here. The variables included will be identified in more detail in Section 10.6.

10.5 Methodology

Simple tabulations of the demographic descriptors and consumer responses are the most common ways used to analyze survey results. Other traditional approaches to find patterns in the data include clustering techniques, e.g., K-means, principal component analysis (PCA), and factor analysis. Multi-dimensional scaling is often used for representing perceived similarities among products or brands.

A common shortcoming of all these methods is the lack of convenient visualization for the decision-maker (e.g., the manager who may be interested in formulating marketing strategies and tactics for consumer segments). In the case of the K-means clustering the set of model vectors that represents the clusters is still a set of points in a high-dimensional space; additional dimensionality-reduction methods are therefore needed for visualization.

Moreover, with traditional clustering methods, the interpretation of clusters is often difficult. Most clustering techniques tend to assign the data to clusters of a particular shape even if there are no actual clusters in the data. If the goal is not just to compress the data but also to make inferences about the cluster structure, it is essential to analyze whether the data set exhibits a clustering tendency.

PCA and factor analysis are standard methods for linear projections of the data onto a much smaller subspace such that the variance of the original data is preserved as much as possible. Indeed, the eigenvalue criterion frequently used is a measure of the variance explained by the model. Yet, as a linear model, factor analysis imposes stringent assumptions on the data and, as the other methods discussed, has severe limitations in visualizing the structure of (nonlinear) data.

Multi-dimensional scaling (MDS) techniques are methods for creating a space in which similarities are represented as distances, given a certain metric. There exists a multitude of MDS approaches: MDS for metric data (where the exact values of the distances between data items are meaningful) and MDS for non-metric data (where the order of the distances between the data items is important). Most MDS methods are computationally intensive and do not construct an explicit mapping function that could be used for mapping new data items. Instead, the projections of all samples are computed in a simultaneous optimization process.

In contrast to the standard methods of representing survey data for segmentation, self-organizing maps, which are widely used in engineering and becoming of increasing value in finance, economics and marketing applications as well, provide easy visualization, impose few assumptions and restrictions, and, like other data-mining techniques, are able to handle large volumes of data in order to isolate patterns and structures in survey data.

SOM is a non-parametric regression technique which is most often used to form a two-dimensional topological representation of the input data. SOM belongs to a class of unsupervised neural network techniques; it is a data-driven approach that extracts relationships from data. The neurons of a SOM represent the general form of the data. Training of neurons through successive presentation of sample data vectors produces an "elastic net" that is stretched to cover the input space of the data. Instead of looking at vast quantities of tabular data or statistics, a SOM provides a map that allows us to visualize an abstraction of the original data. Furthermore, a component plane representation of a SOM provides information about the correlations between data, the division of data in the input space, and the relative distributions of the components. For more details on the SOM approach see Chapter 11.

The SOMs in this paper were created using Viscovery SOMine from Eudaptics Inc. (a software product that is commercially available and is discussed in detail in Chapters 13 and 15). All responses to the CEIBS questionnaire were treated equally, i.e., given the same weight (or priority). The responses were preprocessed by normalization of the columns to variance one so as to obtain a uniform scale for all attributes.

10.6 Major Results

The CEIBS survey of consumers in Beijing and Shanghai, conducted in early 1997, yielded 930 responses of which 436 were from people interviewed in Beijing and 494 from people interviewed in Shanghai. Several records had missing data. When using traditional methods these records with missing data would pose severe problems. SOMs, however, can handle records with missing values.

Since prior research indicated significant differences between consumers in different cities in China, data were analyzed separately for the Beijing and Shanghai samples to avoid a bias toward unjustified generalizations. As shown below, the results justify the separate treatment of Beijing and Shanghai consumers in many respects.

As described earlier, the CEIBS survey gathered the following basic demographic data on consumers in Beijing and Shanghai: gender (male = 1, female = 2); age (student = 1, young = 2, middle-aged = 3, old = 4, senior = 5); income level (none = 0, less than 800 RMB per month = 1, between 800 and 3000 RMB per month = 2, more than 3000 RMB per month = 3); education (primary = 1, secondary = 2, college = 3, more than college = 4); occupation (manual = 1, mental = 2, student = 3, teacher = 4, other = 5); sector of employment (state-owned-enterprise = 1, collectivist = 2, government = 3, foreign enterprise = 4, school = 5, private enterprise = 6, other = 7).

The first four of these variables (gender, age, income level, and education) are progressively ordered. The last two (occupation and sector of employment) do on first sight not appear to be metrically ordered.

In principle one cannot use metric values for inputs to a SOM if the attributes do not represent metric differences. To be used in a SOM these would have to be separated into binary subattributes, e.g. "occupation is manual" becomes zero or one, "occupation is mental" becomes zero or one, etc.

Alternatively, if there is a background scale for each of these two attributes then the coding can be used as presented above. For example, the scoring of "occupation" could be based on a progressive scale of skill and knowledge level: from manual or not skilled, to teacher, the highest skill level. In this case, "other" would need to stand for a skill level higher than "teacher". Note that in Chinese culture to be called a "teacher" has a very special meaning.

Likewise, "employment" could be based on a progressive scale from public to private employment, or from work in less profitable to more profitable environments. Implicitly, there may be a correlation with income since the salaries and bonuses in state-owned enterprises may be a lot less than in private enterprises. In both cases "other" would have to be considered a higher level than its precedent. *In consequence, it is important in the design and coding of a questionnaire to keep in mind the metric ordering of the responses to various attributes. Note that it is very usual to scale items as was done here. This problem is common to all numeric methods of data analysis – and is not peculiar to SOM.* However, other methods often tend to hide this. In view of these limitations we will give only marginal importance to "occupation" and "sector of employment" in the analysis that follows.

Table 10.1 shows the results of using SOM for a basic clustering of this demographic data from the CEIBS survey. The number of clusters was not predetermined. SOM applied to this data mapped into four clusters; the data self-organized itself by minimizing the distance between the input vectors and the original model vectors. Furthermore, the original model vectors adapted during the training process so as to reflect over time the topology of the input data.

The four main clusters in this SOM included: a cluster that matched 73% of the input vectors, a second cluster that matched 14% of the inputs, a third that attracted 8%, and a fourth cluster covering 4% of the input vectors. About 1.4% or 13 responses out of 930 did not fall into any of these four clusters.

The first of these clusters represents mainly a middle-aged group with income between 800 and 3000 RMB per month. Individuals in this cluster have a secondary education and are either involved in collectivist work units or the government. The second cluster represents mainly female students with no income. They already have a secondary education and are in college. The third cluster represents mostly middle-aged and older people who have higher income than the ones in the first group, work for private enterprises, and have a secondary education. Finally, the fourth cluster represents older and more senior people, mostly females, who work for state-owned enterprises or collectivist work units and have a secondary education.

The last column in Table 10.1 shows the weighted average profile of the people

Table 10.1. Basic demographic clusters in the CEIBS survey

	C1	C2	C3	C4	C0	Total
Matching records	677	130	72	36	13	928
Matching records (%)	72.95	14.01	7.76	3.88	1.4	100
Age Group						Sample Average
Mean	3.4	1.1	3.4	4.7	2.5	3.1
Std. deviation	1.2	0.3	1.1	0.7	0.6	
Income						
Mean	2.0	0.1	2.4	1.6	2.5	1.7
Std. deviation	0.7	0.3	0.7	0.7	0.5	
Gender						
Mean	1.5	1.7	1.3	1.8	1.0	1.5
Std. deviation	0.5	0.4	0.5	0.4	0.0	
Education						
Mean	2.5	2.4	2.2	2.1	1.9	2.5
Std. deviation	0.6	0.5	0.6	0.6	0.3	
Occupation						
Mean	2.0	3.1	5.0	6.6	1.9	2.6
Std. deviation	0.8	0.4	0.7	0.8	0.5	
Work Unit						
Mean	2.4	5.0	5.6	1.5	5.5	3.0
Std. deviation	1.7	0.4	1.2	1.1	0.8	

interviewed: they are middle-aged with incomes between 800 and 3000 RMB per month, with secondary education, working mostly for the government.

Our analysis of the preferences, attitudes and behavior of these groups of people is divided into several parts: part 1 examines consumer purchase decisions; part 2 examines media influences; part 3 investigates the importance of product attributes; and part 4 analyzes consumer attitudes regarding brands. Finally, part 5 summarizes difference in lifestyle and dining habits of people in Beijing and Shanghai.

10.6.1 Influences on Purchase Decisions

Figures 10.1 and 10.2 show the SOM results regarding factors influencing consumer purchase decisions in Beijing and Shanghai, respectively. Each of these figures contains both an overall SOM map and the component planes showing the individual factors influencing the purchase decisions. The overall SOM map indicates several distinct clusters – some large (e.g., clusters 1 to 3 in Figure 10.1) and some very small clusters (e.g., clusters 4 to 6 in Figure 10.2). The main difference between the maps for the Beijing and Shanghai samples is that the former shows six clusters and the latter only three (two of which are very small). To determine the optimal cluster size we tuned each of these maps by selecting a cluster significance that gave the best trade-off between number of clusters and their overall significance. In other words, we selected the maximum cluster number before a significant drop in cluster significance occurred.

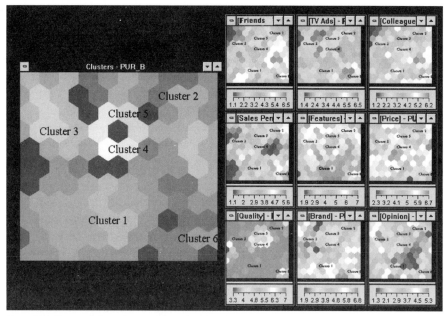

Figure 10.1. Factors influencing purchase decisions by consumers in Beijing, CEIBS survey 1997. The main map shows six clusters; the component planes on the right show the influences of various variables, e.g., advice of friends, TV ads, advice of colleagues, sales person, the product features, price, quality, brand name, the opinion of others. Blue areas in these planes indicate lower values; the red areas indicate the higher values. *(This figure can be seen in color in the Color Plate Section.)*

The component planes in Figures 10.1 and 10.2 show the respective contribution of each variable to the SOM map. The scale at the bottom of each plane goes from the lowest values to the highest values for each variable. For example, in the case of income, the lowest values correspond with the income class that had no income, while the highest values correspond to the highest income class (3000 RMB or more per month).

The component planes thus can be read as follows: in cluster 1 (Figure 10.1), for consumers in Beijing the quality and features of products have very high importance; on the other hand the opinion of friends, colleagues, and sales persons play very little part in their purchase decisions. Consumers in Shanghai are likewise highly influenced by quality and features of products, but appear to attach more importance to the opinion of friends, colleagues and/or the sales person. TV ads play a more important role in Shanghai than in Beijing, as evidenced by the small area of very high values in the TV ads plane in Figure 10.1 compared with Figure 10.2.

While the visual interpretation of the differences between both consumer groups is quite straightforward from these displays, a more detailed analysis of the shape and difference among clusters can be obtained from a "statistical summary." A sample of such an analysis is provide in Table 10.2. The first part of the table shows the purchase decision influences of consumers in Beijing. Note that the SOM produced six clusters (cluster 0 in the last column represents input vectors that did not match closely to the six main clusters identified by the map). The significance of these clusters is shown by the number of matching input vectors (out of a total of 463 responses) as well as the percentage of records in each cluster. Note that in the Beijing sample there are three main clusters with

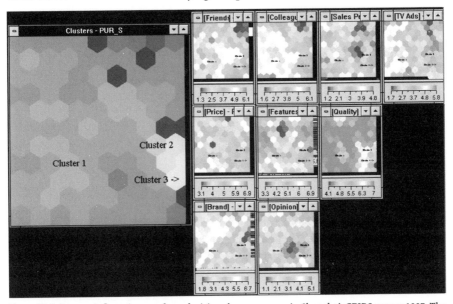

Figure 10.2. Factors influencing purchase decisions by consumers in Shanghai, CEIBS survey 1997. The main map shows three clusters; the component planes on the right show the influences of various variables, e.g., advice of friends, colleagues, sales person, TV ads, price, features, quality, brand name and the opinion of others. Blue areas in these planes indicate lower values; the red areas indicate the higher values. (This figure can be seen in color in the Color Plate Section.)

Table 10.2. Summary of clusters of influences on purchase decisions

Beijing

	C1	C2	C3	C4	C5	C6	C0
Matching records	155	152	68	6	4	5	45
Matching records (%)	35.63	34.94	15.63	1.38	0.92	1.15	10.34
Age level	3.4	2.5	3.9	3.2	2.5	4.0	3.1
Income	2.3	0.9	1.8	1.0	0.8	2.6	2.0
Gender	1.2	1.8	1.2	1.0	1.8	1.4	1.3
Education	2.7	2.2	2.1	1.8	1.8	2.4	2.1
Occupation	2.7	2.5	2.4	2.7	2.5	4.0	3.5
Unit	3.1	3.4	2.2	3.2	4.3	2.8	4.2
Friends	3.7	4.4	1.8	4.2	2.5	5.2	3.4
TV ads	3.8	3.5	2.5	2.7	2.5	4.8	3.9
Colleagues	4.3	4.4	2.0	3.8	4.3	4.8	3.6
Sales person	2.7	3.0	2.2	2.2	2.0	6.2	3.0
Features	5.7	6.3	6.0	4.8	4.3	5.6	5.0
Price	5.0	5.4	4.8	6.5	3.8	6.4	4.7
Quality	6.9	6.9	6.9	6.0	4.3	6.0	5.4
Brand	5.1	4.1	4.3	3.3	3.3	5.2	4.3
Opinion	3.2	3.0	4.4	2.7	3.0	3.8	2.7

Shanghai

	C1	C2	C3	C0
Matching records	443	11	11	28
Matching records (%)	89.86	2.23	2.23	5.68
Age level	3.1	2.7	3.1	2.8
Income	1.8	1.4	2.3	2.1
Gender	1.6	1.1	1.1	1.5
Education	2.5	2.4	2.8	2.6
Occupation	2.5	1.5	2.6	2.4
Unit	2.9	2.8	2.6	3.4
Friends	4.1	4.1	3.0	3.0
TV ads	4.1	3.6	3.6	2.9
Colleagues	4.4	4.4	3.5	3.4
Sales person	3.0	3.0	2.9	2.0
Features	6.1	4.2	5.2	4.7
Price	5.1	4.9	4.5	5.4
Quality	6.8	6.1	4.7	5.9
Brand	4.9	3.7	3.7	4.5
Opinion	3.1	1.5	2.5	3.1

between 15% and 36% of matching records and three small clusters with less than 2% of matching records.

The main differences between the three large clusters in Beijing are as follows:

- Cluster 1 represents mainly middle-aged male consumers (35 to 45 years of age), with 800 to 3000 RMB income per month, who have a college degree or higher and are working for the government. The purchase decisions of this group are mainly influenced by the quality of products, the product features, the brand image and price of a product and to a lesser extend by colleagues, TV ads and the opinion of friends.

- Cluster 2 represents mainly young females between 25 and 35 years that have incomes equal to or less than 800 RMB per month, and have a secondary education or are still students. The purchase decisions of this group are mainly influenced by product quality, product features and price. Friends, colleagues and the opinion of sales persons have a slightly higher influence on this group than on the previous one.

- Cluster 3 represents mainly older people (40 to 49 years of age) who have incomes between 800 and 3000 RMB per month and work in collectivist settings. The purchase decisions of this group are predominately determined by the quality of products and the product features, to a lesser extent by price, brand image, or the opinion of others. A similar group – cluster 6 – whose members work in foreign enterprises, are more influenced by colleagues and sales persons.

Among the consumers surveyed in Shanghai there is only one dominant cluster, representing 89% of the input records. This group represents mainly middle-aged people, aged 30 to 39 years old, who have a college degree. Their purchase decisions are dominated by product features and quality, by brand names and price, and to a lesser extent by friends and colleagues (but more so than in Beijing). The opinion of the sales person is again the least important influence.

A graphic representation of these various influences is shown in Figure 10.3. The x-axis in this figure shows the clusters or groupings of input vectors; the y-axis shows the sample attributes and the z-axis (vertical) displays the relative importance of each attribute in each cluster.

10.6.2 Media Influences

Several question of the CEIBS survey pertained to the influence of the media. Consumers in Beijing and Shanghai were asked how often they watch TV every week, when they watch TV, how often they listen to the radio and when. In addition they were asked to rate on scales ranging from 1 to 7 whether they saw ads as deceptive, entertaining, fun to watch, or informative.

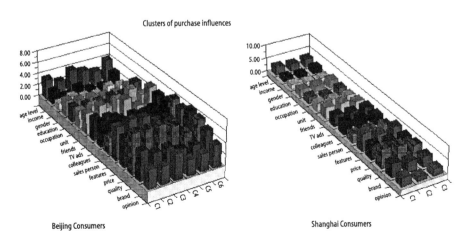

Figure 10.3. Three-dimensional plot of factors influencing purchase decisions of consumers in Beijing.

Table 10.3. TV and radio advertising potential

	TV watching hours/week	TV time[a]	Radio listening hours/week	Radio time[a]
Beijing				
Mean	2.62	4.9	1.68	3.97
St. dev.	1.16	0.74	1.01	2.38
Shanghai				
Mean	2.64	4.94	1.79	3.20
St. dev.	1.08	0.69	0.95	2.28

[a] A value of 3 corresponds to 13:00 to 17:00 hours; a value of 4 corresponds to 17:00 to 21:00 hours and a value closer to 5 corresponds to 21:00 hours or later.

As Table 10.3 shows, there were few difference between consumers in Beijing and Shanghai: both groups watch about 2.6 hours of TV per week, mostly around 9 pm in the evening. On average consumers listen for about 1.7 hours per week to the radio and do this mostly between 5 and 9 pm in the evenings.

Moreover, consumers in both cities think that current advertisements on TV are not very informative (5.39 to 5.58 on a scale of 1 to 7, where 1 stands for very informative and 7 for do not think that ads are informative); they consider some ads as fun to watch (4.12 to 4.32) or entertaining (3.82 to 4.07), yet marginally deceptive (3.1 to 3.25).

The CEIBS questionnaire also asked about "Internet use": 32 out of 436 (or 7.3%) of the consumers interviewed in Beijing currently use the Internet; in Shanghai there were 69 out of 494 or 14% of the respondents using the Internet.

10.6.3 Importance of Product Attributes

Product features, quality and to a lesser extent price and brands were important factors in the purchase decisions of consumers. In this section we analyze these product-related influences in detail. The people interviewed were asked to rate on a scale of 1 to 7, where 7 indicated "very much", whether features, price, functionality/quality and brand image exercised an influence on their decisions. They were also asked about the importance packaging, the retail environment and advertising played in their purchase decisions. Figure 10.4 shows the result of a SOM on the responses to these questions. In particular, we found that

- senior people attach more importance to quality and price than features, packaging, and retail environment. The least important influences on their part were advertising and promotion.
- middle-aged people attributed more importance to product features and quality than to all the other factors.

10.6.4 Consumer Attitudes on Brands

We compared SOMs derived from consumer attitudes regarding brand products. Table 10.4 highlights some of the differences.

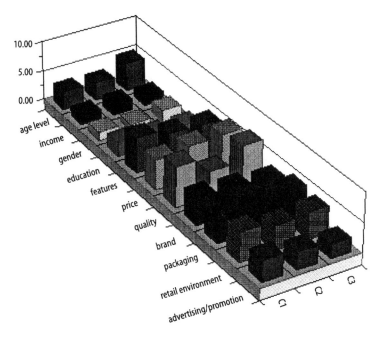

Figure 10.4. Summary results from SOM applied to product- and market-related influences.

The three clusters in Beijing are as follows:

- Cluster 1 represents the young to middle-aged who buy brand products because they think branded products are superior in quality; they like to change brands, particularly of clothes and food, but they do not think that there are major differences between brand products;

- Cluster 2 represents older people who likewise buy brand products primarily because of their superior quality; they do not like to change brands often, and they do not think there are many differences between brands;

- Cluster 3 represents senior people who buy brand products because they are special; they like to change brands of foods and cosmetics; they do not think that there are many difference between brands of appliances, food or cosmetics.

Cluster 1 in the Shanghai map attracted close to 90% of the responses. In this cluster are middle-aged people who buy brand products because they are superior in quality or are special; they like to change brands of clothes and food products, but do not think there are many differences between brand products.

Two other clusters, which are much smaller, represent the young and the senior people, respectively. Like their counterparts in Beijing, the senior people in Shanghai buy brand products because they are special or superior in quality; they like to change brands, primarily of food and personal care products, but they do not think that there are many difference between brands of appliances, food or cosmetics.

Table 10.4. Clusters of brand products

	Beijing			Shanghai		
	C1	C2	C3	C1	C2	C3
Matching records	341	85	3	439	20	9
Matching records (%)	78.39	19.54	0.69	89.05	4.06	1.83
Age level	2.9	3.8	4.0	3.0	2.6	4.6
Income	1.7	1.7	2.0	1.8	1.9	2.4
Gender	1.4	1.5	1.0	1.6	1.9	1.3
Education	2.4	2.2	2.7	2.6	2.2	2.3
Occupation	2.7	2.5	1.7	2.4	2.1	3.3
Unit	3.5	2.2	2.3	3.0	4.0	3.8
I like to buy brand products because they are ... (scale 1 to 6[a])						
Superior–Quality	5.0	5.9	4.7	4.8	3.8	6.1
Special	4.3	4.8	5.3	4.7	3.5	6.4
Fashionable	3.7	4.2	3.3	4.2	3.7	5.6
Not Worth	3.9	4.3	4.3	4.2	3.9	5.7
I change brands often when buying ... (scale 1 to 6[a])						
X–Brand Food	4.5	2.9	7.0	4.7	5.3	6.0
X–Brand Cosmestics	3.3	2.4	6.5	3.3	3.3	4.6
X–Brand Pers care	3.8	2.8	3.3	3.4	2.9	5.7
X–Brand Clothes	4.9	3.9	4.7	5.2	5.5	5.3
X–Brand Appliances	3.6	2.6	5.3	3.8	2.3	4.5
X–Brand Electronics	3.0	1.9	1.3	3.4	3.0	3.9
X–Brand Tobacco	2.8	1.9	1.0	3.0	1.9	2.4
I believe all brands are not very different ... (scale 1 to 6[a])						
S–Brand Food	2.9	4.5	6.3	3.0	4.2	6.8
S–Brand Cosmestics	2.6	4.0	5.5	2.5	5.6	5.5
S–Brand Pers care	3.2	4.7	4.0	2.8	5.8	6.5
S–Brand Clothes	2.9	4.6	5.0	2.9	6.0	6.2
S–Brand Appliances	3.6	5.2	7.0	3.1	5.8	6.7
S–Brand Electronics	2.3	3.6	3.3	2.3	6.3	5.3
S–Brand Tobacco	2.1	3.9	1.0	2.5	4.2	4.8

[a]1 = strongly agree and 6 = strongly disagree

10.6.5 Chinese Lifestyle and Dining Habits

While there are several lifestyle questions in the CEIBS survey that could be analyzed, the remainder of this paper will be restricted to the analysis of one aspect of lifestyle in the survey. The CEIBS survey included several questions on dining habits and food preferences. The specific questions analyzed here include: How many times per week do you go out for business and private dinner? How much money per person do you spend? What restaurants do you prefer (Western, Chinese, no difference)?

The SOM shown in Figure 10.5 shows that an overwhelming majority of the people interviewed go about four to six times per month to a restaurant for private dinner. They also go about two to three times per month to a restaurant for a business dinner. On average they spend about Yuan 41 per person (US$ 4.9) for a private dinner and Yuan 36 per person (US$ 4.3) for a business dinner. They have a small bias towards Chinese restaurants (although the average score between 1.7 and 1.4 would indicate that the preference between Western and Chinese restaurants is almost even).

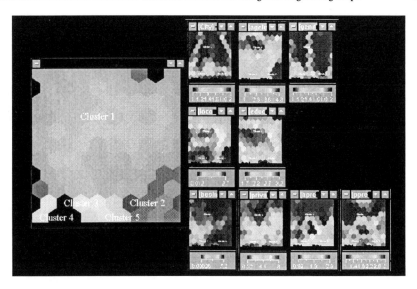

Figure 10.5. Self-organizing map of patterns in dining of Chinese consumers in Beijing and Shanghai, CEIBS survey 1997. The right side of the figure shows five main clusters among over 900 responses on this stratified survey. The component planes show the influences of locations (city: Beijing citizens, blue; Shanghai citizens, red); age (blue is younger, red is senior); gender (blue shows females, red shows males); income (blue shows no income, red shows more affluent); and education (blue shows little, red shows college education) on the number of business and private dinners (where blue areas show lower and red areas show higher frequency of dinners per month), and the cuisine preferences (where blue shows preferences for Chinese restaurants and red shows preference for Western restaurants). *(This figure can be seen in color in the Color Plate Section.)*

Table 10.5. Dining habits

	C1	C2	C3	C5	C4	C0
City	1.64	1.15	1.11	1.11	2.00	1.25
Age level	3.28	1.35	2.74	3.11	2.36	3.51
Income	1.81	0.35	2.37	2.11	2.64	2.19
Gender	1.55	1.65	1.14	1.50	1.36	1.28
Education	2.51	2.30	1.97	1.83	2.91	2.47
Occupation	2.25	2.92	4.09	5.11	2.70	3.56
Unit	2.53	5.18	5.97	6.17	5.10	2.91
Business dinners/week	0.72	0.03	0.32	0.39	6.70	4.03
Yuan/person	36.28	0.45	7.71	17.78	176.36	141.75
Private dinners/week	1.44	1.38	6.06	1.00	2.80	3.21
Yuan/person	41.31	23.08	36.31	28.33	112.73	50.12
Business dinner preferences	1.44	1.33	1.31	1.17	2.18	1.27
Private dinner preferences	1.67	1.35	1.18	1.33	2.18	1.36
Matching records	696.00	89.00	35.00	18.00	11.00	79.00
Matching records (%)	75.00	9.59	3.77	1.94	1.19	8.51

Table 10.5 also shows there are smaller groups in Beijing (cluster 2) who take less business dinners and spend substantially less on dining for both private as well as business purposes. On the other hand, a very small group in Shanghai has six business dinners plus two private dinners per week and spends on average Yuan 176 per person (US $21) for business dinners (five times as much as the majority group). This group also spends three times as much for private dinners (Yuan 112 or US$ 13.5 per person). This small group in Shanghai has a strong preference for Chinese restaurants (average score 2.1).

10.7 Conclusions

In this chapter, we analyzed preferences, attitudes and behavior of Chinese consumers in Beijing and Shanghai based on data from a survey conducted at CEIBS in Shanghai during the spring of 1997.

The key finding of the study is the presence of Beijing and Shanghai segments as well as distinct Beijing and Shanghai subsegments regarding purchase decisions. Aside from these differences, we found that

- The dominant influences on purchase decisions in general are product quality and features; price and brands to a lesser extent; and the opinion of friends, colleagues or sales persons to the least extent.
- TV watching and radio listening occupy a small amount of time; ads on TV are considered somewhat deceptive, but occasionally fun to watch; ads are considered not very informative.
- Product quality and features are more important than other product related influences.
- Most people interviewed buy brand products because they think those products are of superior quality, they like to change brands, particularly of clothes, but they do not think that there are many differences between brands of appliances.
- The dining habits of the majority of people interviewed indicates that they go to restaurants about six times a month for private dinners and about four times a month for business dinners, except for a small group in Shanghai who go more frequently to restaurants and also spend substantially more on dining.

The study illustrates the usefulness of self-organizing maps as a methodology for representing survey data. We used self-organizing maps to represent the data because they are a nonlinear, non-parametric regression technique that yields topological, two-dimensional representations of the data that allow for easy visualization of patterns and structures in the data. Furthermore, SOM analysis is a data-driven approach that has no problems in dealing with incomplete data vectors or in handling non-metric data. SOMs are thus an ideal data analysis instrument for identifying patterns as well as similarities and differences in the marketplace, which can be used to formulate targeted marketing strategies and tactics.

Part 2

Methodology, Tools and Techniques

11 The SOM Methodology

Teuvo Kohonen

Abstract

Conventional statistical methods are able to reveal regularities, trends and structures in raw data. Few methods allow to directly visualize relations between elements in large and complex data sets. In this book we have provided several applications of a method introduced around 1982 called the Self-Organizing Map (SOM). The relations between data items become explicit in the SOM due to a nonlinear projection from a high-dimensional data space onto a two-dimensional display. As demonstrated, this method has been found useful for financial, economic and marketing applications. In this chapter we explain the SOM concepts and methodology, starting from concepts that are supposed to be generally known.

11.1 Regression Principles

Let us start with *regression*, a concept that may be thoroughly familiar to financial analysts. There exist two types of regression that are based on quite different philosophies, and we shall exemplify them by means of a simple line fitting example.

In the *least-squares* (LS) regression, one variable (say x_1) is independent and its values are known or can be selected exactly. The other variable x_2 is supposed to depend on x_1 and we may try to fit a functional form (linear, polynomial, exponential, Gaussian, etc.) to the observation samples by adjusting a small number of parameters of the function. In Figure 11.1a, the sum of squares of the *vertical distances* of the observation points from an assumed straight line is minimized.

Assume now that neither x_1 nor x_2 is known exactly: both are similar stochastic variables. In this case, as illustrated in Figure 11.1b, the line is fitted to the observation points so that the sum of squares of the *orthogonal distances* of the observation points from the line is minimized. This is called the *total least-squares* (TLS) regression.

The LS regression is used in estimation problems, whereas the TLS regression is more often applied to various (especially graphical) smoothing tasks. The SOM

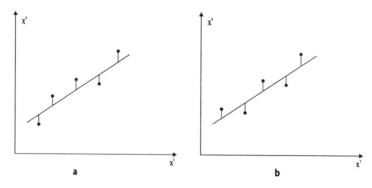

Figure 11.1. **a** Least-squares (LS) regression. **b** Total least-squares (TLS) regression.

method may also be regarded as a special kind of regression, and it is closer to the TLS than the LS regression.

Some mathematicians refer to the so-called principal curve analysis of nonparametric regression, which is a generalization of the TLS regression. That method was invented several years later than the SOM principle, and one can easily show that it is based on the same philosophy as the SOM.

11.2 "Intelligent" Curve Fitting

Let us then discuss a simple operation, which looks like regression and will later turn out to be the *one-dimensional SOM array* principle. In usual regression, the analytical form of some mathematical function is given, and only its parameters are adjusted. A familiar problem thereby is that it is not always easy to guess what form of the mathematical function should be used. A mathematically defined curve may fit part of the points very well while escaping the rest. Attempts to use more "flexible" curves such as a polynomial may lead to *overfitting*: the stray values of the observations do not contain any information (Figure 11.2)

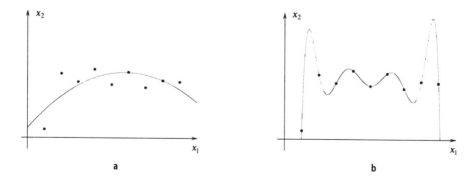

Figure 11.2. **a** Fitting a parabola to stray points. **b** Over fitting a high-order polynomial to the same points.

In "intelligent" curve fitting we do not assume any functional form a priori. The form is automatically determined by the data and certain constraints. However, in the sequel we will not fit any continuous curve to the observation points: we use a set of discrete points called the *nodes* that are linked together. These nodes are represented by sets of real numbers. As discussed earlier, we shall call such ordered sets of numbers *vectors*, or *model vectors*.

Let us consider a two-dimensional observation space (i.e., we have only two variables x_1 and x_2 to be observed) and let $\mathbf{x} = (x_1, x_2)$ be a stochastic *observation vector* from which we have collected a number of samples.

Let us also assume a number of model vectors or nodes, which are similarly represented as the two-dimensional vectors

$$\mathbf{m}_i = (m_{i1}, m_{i2}), \quad i = 1, 2, \ldots, N \ . \tag{1}$$

Let us further define fixed *communication links* between given pairs of nodes (Figure 11.3).

The problem is to fit the nodes $\mathbf{m}_i$ to the samples in an orderly fashion, as if they lie along a flexible curve.

"Regression" in this scheme means that for every sample of $\mathbf{x}$, the closest $\mathbf{m}_i$ called the *winner* $\mathbf{m}_c$ shall first be identified. The "distance" between two vectors is in general computed as the norm of their vectorial difference, where the norm or length (we shall mainly use the so-called Euclidean norm here) of an n-dimensional vector $\mathbf{x}$ is denoted $\|\mathbf{x}\|$ and reads

$$\|\mathbf{x}\| = \sqrt{x_1^2 + x_2^2 + \ldots + x_n^2} \ . \tag{2}$$

Let the "closest" $\mathbf{m}_i$ denoted $\mathbf{m}_c$ be identified by the condition

$$\|\mathbf{x} - \mathbf{m}_c\| = \min_i \{\|\mathbf{x} - \mathbf{m}_i\|\} \ . \tag{3}$$

After that the value of $\mathbf{m}_c$ shall be adjusted, and also its neighbors in the chain must be told through the communication links to adjust their values toward $\mathbf{x}$ in proportion to $\mathbf{x} - \mathbf{m}_i$. This procedure is iterated for all samples of $\mathbf{x}$ again and again.

I am sure that somebody would like to try this kind of fitting. Therefore it will be necessary to give some advice right away. The initial values $\mathbf{m}_i = (m_{i1}, m_{i2})$

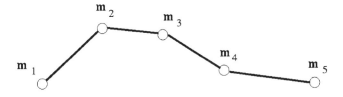

Figure 11.3. A "flexible curve" consisting of nodes that are linked together.

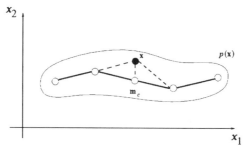

Figure 11.4. An elementary fitting step described in the text. Here $p(\mathbf{x})$ denotes the density function of $\mathbf{x}$.

for the model vectors can be selected at random, and since the $\mathbf{m}_i$ are changing in successive adjustments, we must identify both the observations and the model vectors by the index $t = 1,2,\ldots$. If the samples are regarded as a time series, then t is the discrete-time index of the sample. The fitting operation (Figure 11.4) can be described mathematically as the following series of steps. Let us apply the values $\mathbf{x}(t)$ in succession, and define the corrections on the $\mathbf{m}_i$ as the steps

$$\mathbf{m}_i(t+1) = \mathbf{m}_i(t) + \alpha(t)[\mathbf{x}(t) - \mathbf{m}_i(t)]$$

for those $\mathbf{m}_i$ that lie up to a certain distance along the chain from $\mathbf{m}_c$ and

$$\mathbf{m}_i(t+1) = \mathbf{m}_i(t) \tag{4}$$

for the rest of the nodes. For instance, if $\mathbf{m}_5$ were the winner node, then, if the communication distance were equal to 1, we should adjust nodes $\mathbf{m}_4, \mathbf{m}_5$ and $\mathbf{m}_6$.

The value of $\alpha(t)$ shall be such that we always have $0 < \alpha(t) < 1$, and $\alpha(t)$ must decrease with the steps. In the beginning of the process, $\alpha(t)$ might be selected close to unity. However, an initial value of $\alpha = 0.5$ will also do very well in the first experiments, whereby the process also looks "smoother." The function $\alpha(t)$ according to which a decreases in time is not very crucial; for instance, one can let $\alpha(t)$ go to zero linearly in T steps, where T is at least 100 times the number of nodes, but preferably much larger.

On the other hand, the communication distance along the chain up to which the winner tells its neighbors to make adjustments should be large in the beginning: a rule-of-thumb in this preliminary example is that it should be half of the length of the chain. (In other words, if the winner were the middle node, it would affect the whole chain, but if it were close to one end, it would affect half of the chain.) During the process when $\alpha(t)$ goes to zero, the distance, in discrete steps, might also decrease linearly with time, but its final value should be left to unity, i.e., the closest nodes of the winner should always be updated.

Some people who try this algorithm complain: "The process looks very unstable in the beginning." Why care? If the process will automatically stabilize itself with time, and the final result is good, one should be happy.

It may be intuitively clear that a series of this kind of adjustments will sooner or later lead to a smooth form of the chain, whereas to prove it mathematically has turned out to be extremely difficult (Figure 11.5).

What we are implementing is in fact a *piecewise regression*, and the piece of the chain to be fitted to the observation points is determined in a *decision process*, whereby we always have to identify the closest model vector first.

An example of "intelligent fitting" is given in Figure 11.6.

Instead of referring to fixed links, the *intercommunication strength* h_{ij} between any pair of nodes in the net can be defined as a function of the node indices i and j:

$$h_{ij} = h(i, j) \ . \tag{5}$$

This is also called the *neighborhood function*. During fitting, the winner node can thus also control its neighbor nodes gradually. For instance, in the "intelligent chain" structure we might take for the neighborhood function the Gaussian form

$$h_{ij} = \exp[-(i - j)^2 / 2\sigma^2] \ , \tag{6}$$

where $\sigma = \sigma(t)$ is some suitably chosen, monotonically decreasing function of time (e.g., a linear function that goes to zero). Adjustment of the nodes $\mathbf{m}_i$ around node $\mathbf{m}_c$ will then occur in proportion to h_{ci}. However, h_{ij} is not a function of the vectorial distances between $\mathbf{x}$ and the $\mathbf{m}_i$ or between the $\mathbf{m}_i$, it is a function of indices. In this case all the model points are adjusted at every step, which means a heavier computing load than with the simple communication links, and the adjustment is strongest around the closest model vector.

11.3 The Self-Organizing Map Algorithm

In general, assume that the sets of observation samples constitute n-dimensional real *sample vectors*

$$\mathbf{x}(t) = [x_1(t), x_2(t), \dots, x_n(t)], \tag{7}$$

where t is regarded as the index of the sample $(t = 1, 2, \dots)$. In other words, one observation is an n-tuple of values, and the observation space is n-dimensional:

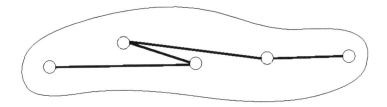

Figure 11.5. It is highly improbable that this kind of zig-zag configuration would result in smoothing.

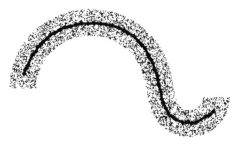

Figure 11.6. A 50-node chain was fitted to the samples (small dots), showing how the form of regression was determined automatically.

similarly, an n-dimensional model vector is associated with each node. As the values of the model vectors are changing in adjustments in response to the samples $\mathbf{x}(t)$, $t = 1,2,\ldots$, that are applied to the fitting algorithm in succession, we must write

$$\mathbf{m}_i(t) = [m_{i1}(t), m_{i2}(t), \ldots, m_{in}(t)] \ . \tag{8}$$

As we shall aim at visualization of the observation space on a two-dimensional "screen," we shall define the neighborhood relationships between the model vectors as if they lie along a *two-dimensional flexible net*. The index of the winner at each step will always be identified first:

$$\left\| \mathbf{x}(t) - \mathbf{m}_c(t) \right\| = \min_i \{ \left\| \mathbf{x}(t) - \mathbf{m}_i(t) \right\| \} \ . \tag{9}$$

After that the $\mathbf{m}_i(t)$ are adjusted:

$$\mathbf{m}_i(t+1) = \mathbf{m}_i(t) + \alpha(t)h_{ci}(t)[\mathbf{x}(t) - \mathbf{m}_i(t)] \ . \tag{10}$$

The factor $\alpha(t)$ $(0 < \alpha(t) < 1)$ is called the *learning-rate factor* and it is decreasing with t. It can be combined with the neighborhood function $h_{ci}(t)$. We shall define h_{ci} more closely below.

This Self-Organizing Map (SOM) algorithm defines fitting of the "intelligent net" to the density function of $\mathbf{x}$, denoted $p(\mathbf{x})$, in the n-dimensional space.

There are still a couple of questions that remain: does the series of values $\{\mathbf{m}_i(t)\}$ converge, and what is the convergence limit? Is it unique? These are very hard problems, and only a very thorough mathematical analysis will give some definite answers to them. Let it suffice to state here that we can be sure of the convergence of the $\mathbf{m}_i(t)$ to reasonably good values $\mathbf{m}_i^*$ if $\alpha(t)$ and $h_{ci}(t)$ are chosen suitably. In Chapter 14 we will advise how this can be done.

The SOM also has a property that may seem a bit confusing, although in practice it does not cause any problems: it is possible to fit the "intelligent net" to the data in any possible symmetrical inversion. However, it is possible to interchange the directions in the display afterwards in any desirable way. Another problem is that there may also exist locally stable configurations of the net that do not represent the best result. We shall see later that it is possible to make several trials and select the best configuration.

Even when the initial values of the $\mathbf{m}_i$ are selected completely randomly, it can be proven that the "net" formed by the $\mathbf{m}_i(t)$ will straighten out in the long run and take its desired form. However, we will point out in Chapter 14 that random initialization is not a good strategy in practice.

Although Eqs. (9) and (10) look superficially simple, one should pay attention to the index c in h_{ci}. It is a variable index, and its value, through (9), depends on $\mathbf{x}$ and all the $\mathbf{m}_i$ discontinuously. So *that* is what makes the algorithm so sophisticated!

11.4 The Neural Network Model of the SOM

All observation vectors $\mathbf{x}(t)$ to which a particular model vector $\mathbf{m}_c$ is closest also select $\mathbf{m}_c$ as the winner, and can be thought to be "mapped" to it. Each model vector then also approximates a whole domain of the observation space. The net formed of the $\mathbf{m}_i$ first takes a form that fits the samples best; after that Eqs. (9) and (10) define a nonlinear projection of the $\mathbf{x}$ space onto the net.

Consider now a regular two-dimensional array (grid) of processor units called the *neurons* (Figure 11.7). For visualization purposes it is best to assume this grid (display) as hexagonal.

With each neuron we thus associate a model vector $\mathbf{m}_i$, i.e., the *coordinates* of a node in the signal space are now represented as a set of real numbers at the corresponding neuron. In traditional neural models the model vector would

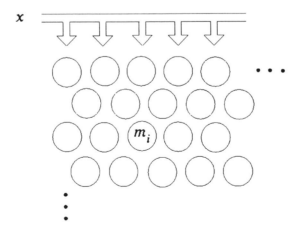

Figure 11.7. Two-dimensional display, the SOM array, where the $\mathbf{m}_i$ are the model vectors at each process unit.

correspond to the input connection weights, but here we need not specify the nature of the $\mathbf{m}_i$.

We also imagine that the sequence of observation samples $\mathbf{x}(t)$ can somehow be "broadcast" to each neuron, one sample at a time. Some kind of computing mechanism (in the simplest case, programmed serial comparison of $\mathbf{x}$ with the $\mathbf{m}_i$) shall first be at work selecting the winner neuron whose $\mathbf{m}_i$ matches best with $\mathbf{x}$. This neuron shall have the index $i = c$.

If we express the location of neuron i in the array by a two-dimensional location vector $\mathbf{r}_i$, then the neighborhood function could have the form

$$h_{ij} = \exp\left[-\left\|\mathbf{r}_i - \mathbf{r}_j\right\|^2 / 2\sigma^2\right]. \tag{11}$$

Even simpler forms of h_{ij} can be used.

Let us recall that the SOM algorithm defines how the $\mathbf{m}_i(t)$ are adapting to the $\mathbf{x}(t)$ values. Let us assume that we have enough samples $\mathbf{x}(t)$, and if not, we reiterate over them, so that there will be enough "learning steps" for the $\mathbf{m}_i(t)$ to converge to acceptable stationary values $\mathbf{m}_i^*$. Any $\mathbf{x}(t)$ thereafter selects one neuron, namely that one, the model vector of which is most similar to $\mathbf{x}(t)$.

The neural network represents the image of the whole observation space: different neurons are sensitized to different observation domains, like a "monitor panel" (described in the Introduction) where different lamps are lit when the system is in different states.

11.5 Labeling the Neurons

We have seen that the nodes of the "intelligent net" seek places in the data space where the samples are concentrated. On the other hand, the SOM array, or the neural-network display, represents the corresponding "coordinates" of the nodes (i.e., the model vectors) at the different neurons. What remains to be done is to label the neurons, or some of them, in order that one can see immediately what the different neurons in the array mean.

In statistical tables, each entry, such as a country or company, is usually described by a unique data set. The names of the entries can then be used as the labels. Assume that the model vectors have converged to acceptable values $\mathbf{m}_i^*$. If $\mathbf{x}(t)$ is the data vector of an entry that has the symbol $s(t)$, and if

$$\left\|\mathbf{x}(t) - \mathbf{m}_c\right\| = \min_i\left\{\left\|\mathbf{x}(t) - \mathbf{m}_i^*\right\|\right\}, \tag{12}$$

we simply label neuron c by $s(t)$.

On the other hand, if the data that describe an entry are stochastic, as the data describing, say, the states of an industrial process are, different measurements of the same entry will generally produce different data vectors $\mathbf{x}(t)$. It is usually possible to give names for certain well-defined process states. Labeling the SOM by these names must then be based on majority voting: if all the vectors $\mathbf{x}(t)$ are input, they will in general label different neurons, and each neuron will be given multiple labels. The majority of labels at a neuron will be selected for the most likely final label.

11.6 The Batch Version of the SOM

Another remark concerns faster algorithms. The incremental regression process defined by (9) and (10) can often be replaced by the following batch computation version, which is significantly faster and does not require specification of any learning-rate factor $\alpha(t)$. Assuming that the convergence to some ordered state is true, the expectation values of $m_i(t+1)$ and $m_i(t)$ for $t \to \infty$ must be equal, even if $h_{ci}(t)$ were then selected nonzero. In other words, in the stationary state we must have

$$\forall i, \ E\left\{h_{ci}(\mathbf{x}-\mathbf{m}_i^*)\right\}=0 \ . \tag{13}$$

In the simplest case $h_{ci}(t)$ was defined as: $h_{ci}=1$ if i belongs to some topological neighborhood set N_c of cell c in the cell array, otherwise $h_{ci}=0$. With this h_{ci} it follows that

$$\mathbf{m}_i^* = \frac{\sum_{V_i} \mathbf{x}(t)}{n(V_i)} \ , \tag{14}$$

where V_i is the set of those $\mathbf{x}(t)$ samples that are able to update vector $\mathbf{m}_i$, and $n(V_i)$ is the number of samples in V_i; in other words, the "winner" node c for each $\mathbf{x}(t) \in V_i$ must belong to the neighborhood set N_i of cell i.

If all observation samples $\mathbf{x}(t)$, $t=1,\dots,N$, are available prior to computations, they can be applied as a batch in the regression, whereby the following computational scheme of (14) can be used:

1. Initialize the model vectors $\mathbf{m}_i$.
2. For each map unit i, collect a list of all those observation samples $\mathbf{x}(t)$, whose most similar model vector belongs to the neighborhood set N_i of node i.
3. Take for each new model vector the mean over the respective list.
4. Repeat from step 2 a few times.

Notice that steps 2 and 3 need less memory if at step 2 we only make lists of the observation samples $\mathbf{x}(t)$ at those units that have been selected for winner, and at step 3 we form the mean over the union of the lists that belong to the neighborhood set N_i of unit i.

11.7 Conclusion

A more detailed explanation of the self-organizing map methodology and its variations can be found in my book *Self-Organizing Maps* (Kohonen, 1997). This brief overview should be sufficient to allow the reader to assimilate the applications presented in this book.

12 Self-Organizing Maps of Large Document Collections

Timo Honkela, Krista Lagus and Samuel Kaski

Abstract

All applications presented in the previous chapters applied self-organizing maps to reducing quantitative, numeric data. This chapter shows how textual information can be treated in a similar way and how self-organizing maps can help in more effective retrieval of information than current search engines. The use of WEBSOM is a novel method for organizing collections of text documents into maps, for browsing and exploring links on the World Wide Web, or for organization of electronic messages or files. Timo Honkela and the team at the Neural Network Center at HUT provide several examples of the use of WEBSOM and many more are available on their website (*http://nodulus.hut.fi/websom/*).

12.1 Introduction

A well-organized library provides several catalogues of the material stored: one by author and one by subject. By looking at the subject catalogue a user can gain an idea of what kinds of books the library has on a particular subject. An ordered collection of links in the World Wide Web is a computerized counterpart of a library. Web directories or ready-made categorization of web links are useful for quick access to information by users. They are also useful if the domain is not very well known, or if the user has only a limited idea of the contents of a domain.

Although a library catalogue may provide invaluable help to find a relevant book or article it may nevertheless be time-consuming because a particular topic may not necessarily fit in the predetermined categories. Searching for information has become easier with the availability of electronic full text databases that have the capability for performing complete text searches. Thus, one can look for documents that contain a specific search expression, for example, a string or a combination of strings.

Used as stand-alone tools, keyword searches are obvious limited. Depending on the size and specificity of the database, and of the quality of the chosen keywords, the search may return hardly any documents or an overwhelming number of them.

In this chapter we describe how SOM can be used to combine both an overall organized view of the document collection and the capability of performing detailed searches. The SOM organizes the documents automatically into maps, where nearby locations usually contain similar documents. The document collection can be explored with the aid of the map view, and content-directed search results provide ideal starting points for exploration of related topics. One may provide keywords or even a whole article for which the system then finds the closest "relatives" on the map. A starting point is always found – there are no null-results or huge outputs.

There are many alternative ways of using the SOM for creating document maps. There are many ways that SOM can be used in information retrieval and textual data mining in general. In this chapter we present only a few approaches including the WEBSOM (Honkela et al., 1996; Kaski et al., 1996; Kohonen et al., 1996; Lagus et al., 1996) [12.01].

12.2 WEBSOM for Document Map Applications

The WEBSOM is a novel method for organizing collections of text documents into maps, and a browsing interface for exploring the maps. The maps are created automatically using the SOM algorithm. With suitable preprocessing, any kinds of text documents can be processed. The name of the system stems from the massive amounts of potentially useful documents that there are available in electronic form in the World Wide Web, including the home pages and Usenet newsgroup articles. The current implementation of the WEBSOM also includes a browsing tool that allows the users to reach the document maps and the organized collection of documents. The browsing tool effectively consists of a set of Web pages. The methodology used to create these examples will be presented in Section 12.3.

By virtue of the SOM algorithm, the documents are positioned on a two-dimensional grid so that nearby locations contain related documents. When the SOM results are transformed into HTML pages, this document collection can be easily explored using a WWW-based browsing environment. During browsing the user may zoom in on any map area by clicking on the map image to view the underlying document space in more detail. The WEBSOM browsing interface is implemented as a set of HTML documents that can be viewed using any graphical WWW browser.

12.2.1 Map of Newsgroup Articles

The Usenet discussion groups on the Internet are somewhat like public bulletin boards or perhaps even more like unmoderated "Letters to the Editor". There are huge numbers of groups, each devoted to a different topic. In the groups people discuss topics of their own interest, ask for information or advice, or offer information, pictures or services. The messages in the groups (usually called articles) are colloquial, mostly rather poorly written, short documents that often contain little topical information. It is not easy to organize them properly. In this sense they form a kind of worst-case scenario for demonstrating organization of text collections or documents.

We retrieved from the Usenet groups a collection of 4600 full-text documents containing approximately 1,200,000 words and organized these with the WEBSOM method. The collection we selected consisted of all the articles that appeared during the latter half of 1995 in the Usenet newsgroup "comp.ai.neural-nets". These 4600 documents were organized on a map of size 24 by 32 nodes. After the map was formed new articles could be added to the map without re-computing it; the goodness of the result naturally depends on how much the topics have changed. At the end of March 1997 the map contained some 12,000 documents.

The WEBSOM browsing interface provides several views of this document collection, at different resolutions. A view of the whole map offers a general overview of the whole document collection (the top view in Figure 12.1, better visible in Figure 12.2). The small dots in the display denote the nodes on the map, and the similarity of the contents of the nearby documents in different parts of the map is indicated by the shades of gray. White denotes a relatively coherent area, whereas dark regions denote major shifts in the content of the documents. This toplevel display may be on to a zoomed map view, to a specific map node, and finally to a single document. These four view levels are shown in Figure 12.1 in the order of increasing detail. The first two levels provide a graphical look at the map display, first at the overall level view and then by providing a closer look at the selected area. As one goes deeper into the details by moving to the next level the contents of an individual node are revealed, and finally a document is seen (the view at the bottom of Figure 12.1).

In a typical browsing session, the user may start from the overall map view and proceed to examine further a specific area, perhaps later gradually wandering to nearby areas containing related information. After finding a particularly interesting map node the user can use it as a "document bin" which can be bookmarked and checked regularly to see if interesting new articles have arrived.

Let us now examine a specific area in an organized map. As shown in Figure 12.2, WEBSOM positioned articles related to financial issues and fuzzy logic in the middle of the right edge of the map. A closer look at the contents of a few nodes (Figure 12.3) shows that a continuum can be found between discussions of economic applications on the one hand and fuzzy set theory and neural networks methodology on the other. The strictly economy-related node, the titles of which are shown in box 5 in Figure 12.3, contains articles related to financial applications of neural networks, specifically bankruptcy predictions. Moving up on the map, the nodes begin to have articles regarding fuzzy logic in addition to economic issues (inset 3). Finally, in node 2 there are no more articles on economic issues, and fuzzy neural nets seem to have taken over the discussion.

The example shown in Figure 12.3 demonstrates the principles according to which WEBSOM organizes documents. In general the map tries to model the "document space", the space in which the contents of the documents in the document collection can be represented so that nearby locations would represent similar documents. The reason why the articles on neuro-fuzzy systems and economic applications appear right next to each other, and even overlap on the map, is that many of the articles mention both subjects. There is thus an association between these two topics which might a priori seem independent.

In addition to explorations, WEBSOM may be used for content-directed

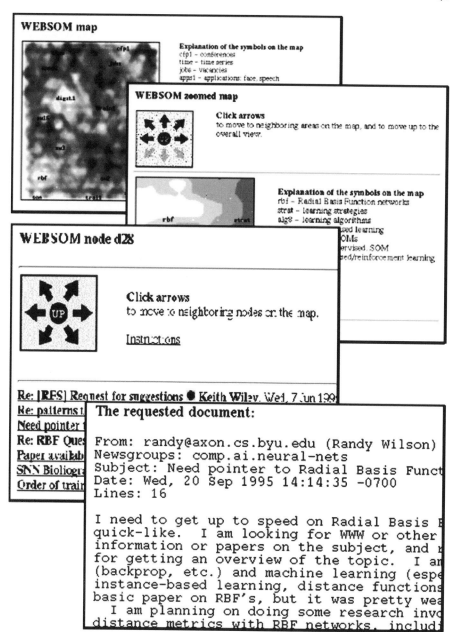

Figure 12.1. The four different view levels of the WEBSOM browsing interface: the whole map, the zoomed map, the map node, and the document view, presented in the order of increasing detail. Moving between the levels or to neighboring areas on the same level is done by mouse clicks on the images or on the document links. Once an interesting area has been found on the map, exploring the related documents in the neighboring areas is simple. This can be contrasted with traditional information retrieval techniques where the users often cannot know whether there is a considerable number of relevant documents just "outside" their search results.

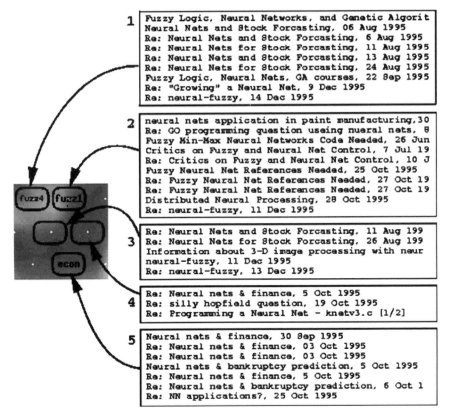

Figure 12.2. Schematic illustration of a WEBSOM document map used as an information filtering tool. The circle denotes the user's interest area. The symbols inside the circle denote documents that would be selected by the system automatically. Those documents could, for example, be instances of interesting electronic mail or articles from a news supplier. The visual map display can also be used to aid in noticing and checking for related documents in nearby areas.

document search. Any new document, for example a free-text query or an otherwise interesting document, may be mapped onto the document display precisely like any of the previous articles. The position of the new document on the document map provides a starting point for exploring related documents in nearby areas. Furthermore, relevant areas on the document map can be used as "mailboxes" into which target information is automatically gathered (Figure 12.2).

12.2.2 Map of Conference Abstracts

Another example of the application of WEBSOM is one involving the abstracts of articles accepted for presentation in the Workshop on Self-Organizing Maps held in Helsinki in 1997 (WSOM'97). These abstracts were organized using a simplified form of the WEBSOM method. The resulting map offers a visual overview of the

Incoming documents

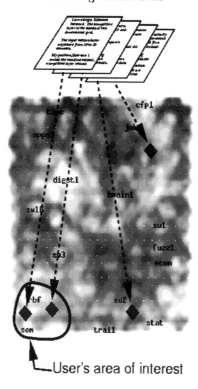

Figure 12.3. An area of the map of neural network articles that contains articles on economic applications of neural networks and neuro-fuzzy systems in neighboring map nodes.

workshop contents. The interactive version of the map can be reached via the WWW address *http://websom.hut.fi/websom/*.

The document collection contained 58 short abstracts, with on average 106 words per abstract, excluding articles ("a", "the", "an") and including the title of the abstract. An abstract of a scientific article may typically introduce the main components of the methodology used, discuss the relations of the work to earlier studies, mention performed simulations and comparisons, and portray an application area of the method or perhaps even several potential applications. Thus, each article may have common subjects of discussion with many other articles. In other words, most of the articles will be related to many different kinds of articles and therefore the actual local dimensionality of the document space may be considerably greater in the case of scientific abstracts than with most Usenet newsgroup articles.

When organizing the articles all of the relations cannot be taken into account. The method must concentrate on visualizing the most salient ones. Furthermore, such grounds for organization may be different in different areas of the document space.

The map of the conference abstracts is shown in Figure 12.4, labeled by manually selected words. After viewing the node contents, many rather interesting small clusters seem to emerge, for example the text processing, finance, and speech recognition areas. In addition, a more extensive area of related theoretical issues can be found under and around the "topology" and "organization" labels. A number of papers discussing a specific application of SOM to the analysis of data sets or to monitoring processes, as well as articles on SOM-based tools, can be found near the visualization label.

Exploration of the map leads to the impression that some areas were organized more based on similarity in theoretical issues (e.g., an area where the topology preservation and organization measures are discussed), whereas in other areas the common factors between abstracts were more often related to the application area, such as in the text-processing area. In still other map areas the abstracts seemed to have both kind of commonalities.

Speech recognition is one example; three of the four abstracts in three nearby nodes mention speech recognition, whereas the fourth abstract seems to be drawn to one of the nodes because it discusses the same theoretical issues as the other abstracts in the same node (i.e., Gaussian mixtures).

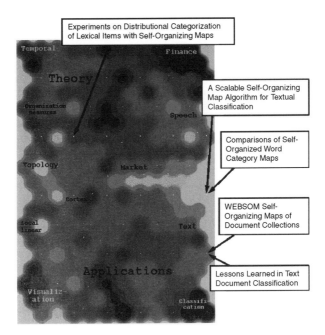

Figure 12.4. The WSOM'97 abstract map. The map has been labeled manually, and the titles of the workshop session "Text and document maps" are shown in the boxes. One of the articles, the one whose title is shown in the topmost box, has found a place quite apart from the other articles in the session. When reading the abstracts the reason becomes evident: the topmost abstract concentrates on linguistic and grammatical aspects and therefore also uses rather different language than the others which have a more practical approach to organizing real-world text and documents.

The labels "Theory" and "Applications" have been included to indicate the perceived grounds of map organization: on the upper left side, abstracts more often seem to be found together because of their theoretical commonalities, whereas lower and right-side areas (except a single node in the lower right corner) appear to be organized mainly according to application considerations.

12.2.3 Other Applications

The idea of using SOM in the exploration of document collections became more widely known in 1990 after the publication of an article by Doszkocs et al. (1990). They quoted Kohonen who said that SOM "is able to represent rather complicated hierarchical relations of high-dimensional spaces in a two-dimensional display." They concluded that document spaces are certainly such high-dimensional spaces. Concrete experiments based on the idea were first published by Lin et al. (1991) and Scholtes (1991), and related approaches have been published, e.g., by Chen et al. (1996) and Merkl (1993).

In the World Wide Web, one obvious application of the document maps is the ordering of home pages, or in fact any available documents. For example, an information provider may organize its material for easier public use. A company may also use the WEBSOM in an Intranet application. Often, the necessary information needed in a particular task can be found inside the organization but the means for finding it are needed. Also electronic mail messages could be automatically positioned on a suitable map according to personal interests. Relevant areas and single nodes on the map can be used as "mailboxes" into which specified information is automatically gathered. The method could also be used to organize official letters, personal files, library collections, and corporate full-text databases. A document map or a collection of maps provides an integral solution.

Administrative or legal documents may be difficult to locate by traditional information retrieval methods because of the specialized terminologies used. For instance, the product developers of a company are likely to express themselves in different terms and paraphrases than the marketing staff. The category-based and redundantly encoded approach of the WEBSOM method is expected to diminish the terminology problem.

12.3 Document Map Creation

12.3.1 Document Encoding

Given the above examples we will now explain the methodology for creating document maps.

The documents must be encoded before they can be organized. The ordering of the document maps depends on the chosen document-encoding scheme. Therefore the encoding should retain the relationships between the contents of the documents, while still being computationally efficient. When aiming at an organization that is based on the topical content of the documents it is useful to discard information that is irrelevant in distinguishing different topic areas. Examples of such information might be exact synonyms and "connector words" used by the authors of the documents.

Perhaps the most straightforward approach would be to limit to encoding only the titles of the articles instead of the full text, hoping that the titles summarize the essentials of the content. However, in most cases this choice is unsatisfactory because the title often gives a very limited or even a misleading view of the contents of an article. Statistical variation is better overcome if more information can be utilized for each article.

A straightforward document encoding is achieved by representing each document as a histogram in some "relevant" vocabulary. In other words, the document is represented by a real-number vector, where each word of the vocabulary corresponds to a component. The value of a component tells how many times the corresponding word appeared in the document, or is a suitable function of this frequency of occurrence. This encoding method is often referred to as the vector space model (Salton and McGill, 1983). Another version of this approach is to calculate occurrences of several consecutive words or letters, n-grams, instead of single words. Clearly, if the vocabulary is large, the document vectors are dimensionally too large for practical computations.

For computational reasons the vector space model used is suitable only in situations where the relevant vocabulary is for some reason very small, such as when the document collection is very tightly concentrated around some topic or when the relevant vocabulary is chosen manually. The vector space model may also be used when the document collection is small and the computational burden is thus not so big.

Another problem with vocabulary-based representations, besides the computational burden, is that they do not take into account the synonymy, or in general the fact that some pairs of words are more similar or more related than others. In the vector space model each word is equally distant from any other word. The problem becomes emphasized when the material to be organized has been written by different authors who may favor different choices of words. In a useful full-text analysis method synonymous expressions should therefore be encoded similarly, and in a computationally effective manner.

12.3.2 Two-level Architecture of the WEBSOM Method

WEBSOM is an explorative full-text information retrieval method that is based on applying the SOM in two processing stages on the document collection.

First, word category maps (also called self-organizing semantic maps; see Ritter and Kohonen, 1989) are formed. They are SOMs that have been organized according to word similarities. A simple way of measuring the similarity of words is to compare their average short contexts. Each word in the sequence of words is first represented by a fixed n-dimensional real vector with random-number components. The averaged context vector of a word then consists of the estimate of the expected value of the representations of the words that have co-occurred with it in a text corpus. The word category map is a SOM that has organized the words based on their context vectors.

The map is calibrated after the training process by inputting the word context instances once again to the word category map and labeling the best-matching nodes according to the words. Usually, interrelated words that have similar contexts appear close to each other on the map (Figure 12.5). Each node may become labeled by several symbols, often synonymous or belonging to the same closed category, thus forming "word categories" in the nodes.

To reduce the computational load the words that occur only a few times in the whole text corpus were neglected before the analysis and treated as empty slots. In order to emphasize the subject matter of the articles and to reduce erratic variations caused by the different discussion styles, common words that are not supposed to discriminate any discussion topics were discarded from the vocabulary.

In the second level of analysis the documents are encoded by mapping their text, word by word, onto the word category map, whereby a histogram of the "hits" on it is formed. To speed up the computation, the positions of the word labels on the word category map may be looked up by hash coding. To reduce the sensitivity of the histogram to small variations in the document content, the histograms are "blurred" on the two-dimensional map using a Gaussian convolution kernel. The document map is then formed with the SOM algorithm using the histograms as "fingerprints" of the documents.

The document map has been found to reflect relations between documents; similar documents tend to occur near each other on the map. Not all nodes may be well focused on one subject only, however. While most discussions seem to be confined into rather small areas on the map, the discussions may also overlap.

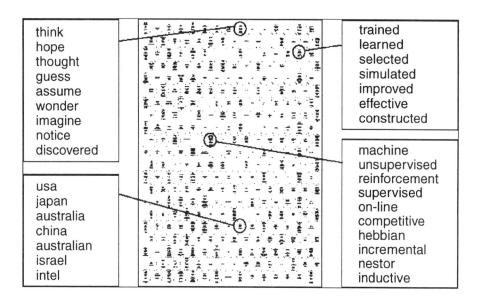

Figure 12.5. A word category map. The contents of four map nodes are shown in the boxes.

12.3.3 Other Approaches

There exist also other document encoding schemes which try to take into account the relations of different words in the encoding. In the latent semantic indexing (LSI) (Deerwester et al., 1990) each word is encoded with a vector that reflects the co-occurrence of the words in the same documents. Words that have occurred often in the same documents will attain similar codes, and the documents will be encoded as weighted sums of the codes of their words. The codes will be formed by a linear method that is based on matrix algebra.

Another related method has been used in the HNC's MatchPlus system (Gallant, 1991). Each word is encoded with a "context vector" and each document is encoded as a weighted sum of the context vectors of its words. The context vectors can be formed by means of linear algebra somewhat like in the LSI (Hecht-Nielsen, 1994), or they can be formed by judging manually how similar each word is to a set of "basis words" (Gallant et al., 1992). Each dimension of the context vectors then represents the manually evaluated similarity of meaning between the word to be encoded and one of the basis words.

To achieve maximal generality of the results we have used only the textual content of the documents for the encoding in the WEBSOM. It would also be possible to incorporate some prior information about the words, such as information obtained using a thesaurus. If the text is written in the form of a hypertext it is possible to incorporate information about the linkages between the documents (Girardin, 1995). Likewise, in scientific texts it is possible to include information about references to other documents.

12.4 Conclusions

Using Self-Organizing Maps for analyzing and visualizing large document collections is a novel approach for information retrieval and data mining. The potential of the method is based on its capability to generate overall views of document collections. The views can be used for exploration, and the map can also be used for associative search. Moreover, the word category maps help in finding a "common language" for different authors and for the users of the information. It is not necessary to use exactly the same words and phrases as is required in keyword-based systems.

The widespread commercial use of the WEBSOM naturally requires system integration in which the new technology is adopted to be used alongside the traditional word processing and information retrieval systems. In an optimistic scenario, the map becomes familiar for all computer users who need to deal with large text document collections. It remains to be seen whether this kind of development leads to inclusion of neural networks technology into operating systems and even commonly used hardware. One could then replace "c" by "s" in the word "computer"...

13 Software Tools for Self-Organizing Maps

Guido Deboeck[1]

Abstract

Data analysis, clustering and visualization with SOM is commonly done with (a) public domain software, (b) self-coded software or (c) commercial software packages. In particular, there is an increasing number of commercial, off-the-shelf, user-friendly software tools that are becoming more and more sophisticated. This chapter contains a brief overview of several public domain software tools as well as a list of commercially available neural network tools that contain a self-organizing map capability.

13.1 Overview of Available Tools

Most applications discussed in this book were done with *public domain software* or software coded by the authors themselves. Several authors used SOM_PAK and extended it with special programs for pre- and post-processing of data to meet the specific needs of their study. The selection of mutual funds (in Chapter 3) and the mapping of investment opportunities in emerging markets (in Chapter 6) was done with SOM_PAK. Carlson and Tulkki used SOM_PAK, but combined it with GIS and other programs to do the analysis of real estate markets. SOM_PAK is particularly suitable for scientific work and applications on UNIX machines; an MS-DOS version is available; however, SOM_PAK does not run under Windows 95 or Windows NT. SOM_PAK will be discussed in Section 13.2.

SOM software on other environments includes the SOM Toolbox. The SOM Toolbox runs on MatLab version 5 and requires no knowledge of the C programming language. This toolbox extends the use of SOM to environments that are becoming standards in financial institutions as well as for training in colleges. While relatively new at the time of writing (September 1997), the SOM Toolbox still needs to be tested and evaluated in terms of its performance and capabilities for handling large applications. The SOM Toolbox will be discussed in Section 13.3.

---

[1] This overview of software tools has been assembled from material produced by Teuvo Kohonen and the staff of the Neural Network Research Center and the Laboratory of Computer and Information Sciences at HUT. It has also greatly benefited from the inputs of Dr Samuel Kaski and Dr Gerhard Kranner.

Another public domain software package is Nenet, a 32-bit Windows 95 and Windows NT 4.0 application, designed to facilitate the use of the SOM algorithm. Nenet can be obtained from the Nenet web site at: *http://www.hut.fi/~jpronkko/nenet.html.* Nenet v1.0 is user-friendly, has good visualization capabilities, and a GUI allowing efficient control of the SOM parameters and visualization methods. Nenet also includes some more exotic and involved features, especially in the area of visualization.

These three public domain SOM packages have been developed for different purposes. Samuel Kaski defined the differences as follows: "SOM_Pak is a professional package for possibly large and computationally intensive studies. NeNet is a package that is easy to use but it suits only small-scale problems. SOM Toolbox is a compromise between these two: it provides flexibility and ease of use but is not as efficient computationally as SOM_Pak and cannot tackle as large problems" (personal correspondence).

Several *off-the shelf commercial software* tools for implementing SOM are available on the market. Not all of them perform in the same way or have the same effectiveness as the public domain software tools. However, the commercially available software packages are designed for running on commonly used environments and platforms and have better imbedded capabilities for pre- and post-processing of the data.

A list of available SOM packages can be found in Box 1. This list was derived from the FAQ on neural networks on the web. It contains only those neural network packages that have SOM capability (see *ftp://ftp.sas.com/pub/neural/FAQ.html*). In the Appendix we provide a short description of available commercial software tools with self-organizing map capability.

Box 1: Commercially available SOM packages

1. SAS Neural Network Application
2. Professional II+ from NeuralWorks
3. MATLAB Neural Network Toolbox
4. NeuroShell2/NeuroWindows
5. NeuroSolutions v3.0
6. NeuroLab, A Neural Network Library
7. havFmNet++
8. Neural Connection
9. Trajan 2.1 Neural Network Simulator
10. Viscovery SOMine

User-friendly software running on commonly used operating environments is a prerequisite for wider adoption of the SOM approach, especially in finance and economics. In financial institutions and economic research centers there are few people who still have time for programming in C, for complex software installations, or non-standard environments. Most applications used in these organizations are off-the-shelf packages, with good user interfaces, easy to install, and easy to use for real-world applications. Most of real-world applications use large data sets but not really huge ones (e.g. for most modeling of financial markets one seldom uses over 1000 trading days). Furthermore, computational efficiency is the least important criterion when it comes to development of financial, economic or marketing models that save costs or are profitable.

13.2 SOM_PAK: The SOM Program Package

An extensive program package, SOM_PAK, which facilitates the main steps in SOM including the selection of the map size and format, proper initialization, monitoring of the computational process, and analysis and interpretation of the resulting mapping, was developed a few years ago. This is a public domain software usable for scientific work. The program code is available on the World Wide Web, at

http://nucleus.hut.fi/nnrc/nnrc-programs.html.

All programs and full documentation of this package are stored in the directory som_pak.
The main program groups included in this directory are:

• Initialization Programs
The initialization programs initialize the reference vectors.

randinit – This program initializes the reference vectors to random values. The vector components are set to random values that are evenly distributed in the area of corresponding data vector components. The size of the map is given by defining the x-dimension (-xdim) and the y-dimension (-ydim) of the map. The topology of the map is defined with option (-topol) and is either hexagonal (hexa) or rectangular (rect). The neighborhood function is defined with option (-neigh) and is either step function (bubble) or Gaussian (gaussian).

```
randinit -xdim 16 -ydim 12 -din file.dat -cout file.cod
-neigh bubble -topol hexa
```

lininit – This program initializes the reference vectors in an orderly fashion along a two-dimensional subspace spanned by the two principal eigenvectors of the input data vectors.

```
lininit -xdim 16 -ydim 12 -din file.dat -cout file.cod
-neigh bubble -topol hexa
```

• Training Programs

vsom – This program trains the reference vectors using the self-organizing map algorithm. The topology type and the neighborhood function defined in the initialization phase are used throughout the training. The program finds the best-matching unit for each input sample vector and updates those units in the neighborhood of it according to the selected neighborhood function.

The initial value of the learning rate is defined and will decrease linearly to zero by the end of training. The initial value of the neighborhood radius is also defined and it will decrease linearly to one during training (in the end only the nearest neighbors are trained). If the qualifier parameters (-fixed and -weight) are given a value greater than zero, the corresponding definitions in the pattern vector file are used. The learning rate function a can be defined using the option -alpha_type. Possible choices are linear and inverse_t. The linear function is defined as $a(t) = a(0) (1.0 - t / rlen)$ and the inverse-time type function as $a(t) = a(0) C / (C + t)$ to compute $a(t)$ for an iteration step t. In the package the constant C is defined as $C = rlen / 100.0$.

```
vsom -din file.dat -cin file1.cod -cout file2.cod -rlen
10000 -alpha 0.03 -radius 10 [-fixed 1] [-weights 1] [-
alpha_type linear] [-snapinterval 200] [-snapfile file.snap]
```

Notice that the degree of forcing data into specified map units can be controlled by alternating "fixed" and "nonfixed" training cycles.

· Quantization Accuracy Program

qerror – The average quantization error is evaluated. For each input sample vector the best-matching unit in the map is searched for and the average of the respective quantization errors is returned.

```
qerror -din file.dat -cin file.cod [-qetype 1] [-radius 2]
```

It is possible to compute a weighted quantization error $S h_c \|x - m_c\|^2$ for each input sample and average these over the data files. If option -qetype is given a value greater than zero, then a weighted quantization error is used. Option -radius can be used to define the neighborhood radius for the weighting, default value for that is 1.0.

· Monitoring Programs

visual – This program generates a list of coordinates corresponding to the best-matching unit in the map for each data sample in the data file. It also gives the individual quantization errors made and the class labels of the best matching units if the latter have been defined. The program will store the three-dimensional image points (coordinate values and the quantization error) in a similar fashion as the input data entries are stored. If a input vector consists of missing components only, the program will skip the vector. If option -noskip is given the program will indicate the existence of such a line by saving line
-1 -1 -1.0 EMPTY_LINE
as a result.

```
visual -din file.dat -cin file.cod -dout file.vis [-noskip 1]
```

sammon – Generates the Sammon mapping (Sammon, 1969) from n-dimensional input vectors to 2-dimensional points on a plane whereby the distances between the image vectors tend to approximate to Euclidean distances of the input vectors. If option -eps is given an Encapsulated PostScript (eps) image of the result is produced. The name of the eps-file is generated by using the output file basename (up to the last dot in the name) and adding the ending _sa.eps to the output filename. If option -ps is given a PostScript image of the result is produced. The name of the ps-file is generated by using the output file basename (up to the last dot in the name) and adding the ending _sa.ps to the output filename. In the following example, if the option -eps 1 is given, an eps file named file_sa.eps is generated.

```
sammon -cin file.cod -cout file.sam -rlen 100 [-eps 1] [-ps 1]
```

planes – This program generates an Encapsulated PostScript code from one selected component plane (specified by the parameter -plane) of the map imaging the values of the components using gray levels. If the parameter given is zero, then all planes are converted. If the input data file is also given, the trajectory formed of the best-matching units is also converted to a separate file. The eps files are named using the map basename (up to the last dot in the name) and adding _px.eps (where x is replaced by the plane index, starting from one) to it. The trajectory file is named accordingly, adding _tr.eps to the basename. If the -ps option is given a PostScript code is generated instead and the produced files are named by replacing .eps by .ps.

In the following example, a file named file_p1.eps is generated containing the plane image. If the -din option is given, another file file_tr.eps is generated containing the trajectory. If the -ps option is given then the produced file is named file_p1.ps.

```
planes -cin file.cod [-plane 1] [-din file.dat] [-ps 1]
```

umat – This program generates an Encapsulated PostScript code to visualize the distances between reference vectors of neighboring map units using gray levels. The display method has been described by Ultsch (1993b). The eps file is named using the map basename (up to the last dot in the name) and adding .eps to it.

If the -average option is given the gray levels of the image are spatially filtered by averaging, and if the -median option is given median filtering is used. If the -ps option is given a PostScript code is generated instead and .ps ending is used in the filename.

In the following example, a file named file.eps is generated containing the image.

```
umat -cin file.cod [-average 1] [-median 1] [-ps 1]
```

- Other Programs

vcal – This program labels the map units according to the samples in the input data file. The best-matching unit in the map corresponding to each data vector is searched for. The map units are then labeled according to the majority of labels "hitting" a particular map unit. The units that get no "hits" are left unlabeled. By giving the option -numlabs one can select the maximum number of labels saved for each codebook vector. The default value is one.

```
vcal -din file.dat -cin file.cod -cout file.cod [-numlabs 2]
```

Some more advanced features have been added into the SOM_PAK program package in Version 3.1. These features are intended to ease the usage of the package by offering ways to use, for example, compressed data files directly and to save snapshots of the map during the training run.

The advanced features include: (1) buffered loading (the whole data file need not be loaded into memory at once); (2) reading and writing of compressed files and piped standard input and output – this facilitates the piping of the commands; (3) saving of snapshots of the codebook during teaching; (4) environment variables.

vfind – The easiest way to use the SOM_PAK programs is to run the **vfind** program, which searches for good mappings by automatically repeating different random initializing and training procedures and their testing several times. The criterion of a good mapping is a low quantization error. The **vfind** program asks all the required arguments interactively. The user only needs to start the program without any parameters (except that the verbose parameter (-v), the learning rate function type parameter (-alpha_type), the quantization error type parameter (-qetype) and the qualifier parameters (-fixed and -weights) can be given).

13.3 SOM: A MatLab Toolbox

The SOM Toolbox is a public domain program package for undertaking SOM applications using MatLab 5 computing environment from Mathworks Inc. For availability of this public domain program package, please consult the URL

http://www.cis.hut.fi/projects/somtoolbox/

As MatLab has steadily gained in popularity as "the" language of scientific computing, there is a need for a MatLab implementation of the algorithms described in SOM_PAK.[2] MatLab is also better-suited for fast prototyping and customizing than the C language used in SOM_PAK. Furthermore, MatLab 5 employs a high-level programming language with strong support for graphics and visualization. The SOM Toolbox takes full advantage of the strengths of MatLab. The SOM Toolbox is not just a SOM_PAK rewritten for MatLab but a new set of programs. The SOM Toolbox is intended to complement the SOM_PAK, not to replace it.

13.3.1 SOM Toolbox Features

Highlights of the SOM Toolbox include the following:

- Graphical user interface (GUI): even the more advanced features of the Toolbox are simple to use – the GUI first guides through the   initialization and training procedures of the map, suggesting sensibly chosen default values for most parameters, then it offers a variety of different methods to visualize the data on the trained map.

---

[2] MatLab Neural Networks Toolbox includes a SOM but the functionality of the original MatLab Toolbox on Neural Networks is very limited.

- Data preprocessing tools: two basic data preprocessing tools – variance normalization and histogram equalization – are included in the package.
- Modular programming style: the Toolbox code utilizes MatLab structures and the functions are constructed in a modular manner, which makes it convenient to tailor the code for each user's specific needs.
- Component weights and names: the input vector components may be given different weights according to their relative importance, and the components can be given names to make the figures easier to read.
- Batch or sequential training: in data analysis applications, the speed of training may be considerably improved by using the batch version.
- Map dimensions: maps may be N-dimensional – although visualization is problematic when $N > 2$.
- Advanced graphics: building on MatLab's strong graphics capabilities, attractive figures can be easily produced. In addition to these, most features found in the SOM_PAK are in some form available also in the SOM Toolbox. Specifically, the only features of the SOM_PAK not implemented in the SOM Toolbox are the "fixed-point qualifiers" for forcing some input vectors to specified locations on the map, and the so-called advanced features: buffered loading of data, data redirection and compression, snapshots of the codebook during learning, and using environment variables.

The SOM_PAK files can also be accessed with the Toolbox, so it is possible to first train the map with the SOM_PAK and then use the Toolbox for map visualization. The SOM Toolbox is programmed using the "plain" MatLab 5 language, so no other Toolboxes are required.

13.3.2 Comparison between SOM_PAK and SOM Toolbox

SOM_PAK and SOM Toolbox each have their relative strengths and weaknesses, some of which are:

- System requirements: Toolbox requires MatLab 5 (which is available for most common operating systems) and puts a heavier load on the memory size and processor speed; SOM_PAK requires only a (ANSI-C compliant) C compiler, and in MS-DOS, not even that as a precompiled version exists.
- Speed of operation: with Toolbox, the map training is about one order of magnitude slower than with SOM_PAK.
- Customization: the MatLab code of Toolbox is much easier to modify – requiring less expertise and time – than the C code of SOM_PAK.
- User interface: Toolbox may be used from the MatLab command line or, on a windowing operating system, with a graphical user interface (GUI; see the examples illustrated in Figures 13.1 to 13.3); SOM_PAK may be used only from the shell command line.
- Figure formats: Toolbox, being based on MatLab, can produce and print figures in most commonly used formats, such as BMP, PCX, PPM, JPEG, TIFF, and PostScript; SOM_PAK produces figures in PostScript only.

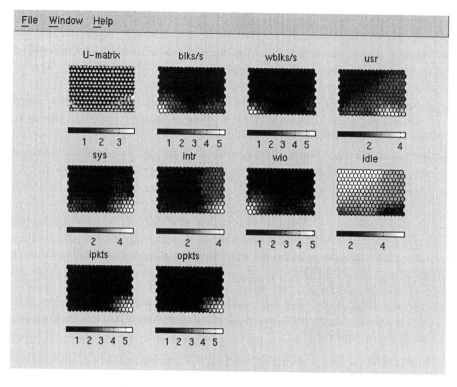

Figure 13.1. Sample windows from SOM Toolbox.

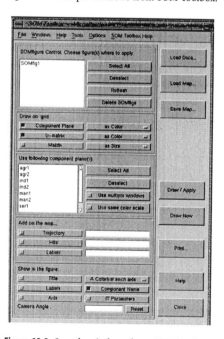

Figure 13.2. Sample windows from SOM Toolbox.

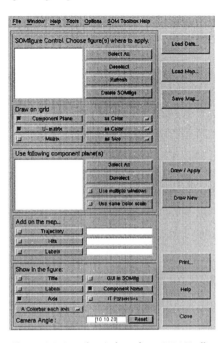

Figure 13.3. Sample windows from SOM Toolbox.

13.4 Viscovery SOMine Lite: User-friendly SOM, at the Edge of Visualization

One of the commercially available software packages is Viscovery SOMine (VS), a product of Eudaptics GmbH in Austria. We have tested this software extensively and have applied it for the example in the Introduction, the analysis of the Chinese consumer survey in Chapter 10, and the credit risk analysis in Chapter 15.

We found Viscovery user-friendly, flexible, and powerful. By powerful we mean that the software provides a substantial number of the features which are essential to expedite financial, economic and marketing applications. Viscovery forms a bridge between state-of-the-art algorithms and the need for a user-friendly, easy-to-use tool. It is a package for advanced analysis and monitoring of numerical data sets and comes in two flavors: SOMine Lite and SOMine Pro. The Pro flavor comprises the Lite version but has enhanced clustering options, dependency analysis and an OLE interface (eventually it will also have an SQL and DB2 interface).

Viscovery is based on the SOM concept and algorithm; it employs an advanced variant of unsupervised neural networks, i.e. Kohonen's Batch-SOM. In contrast to the standard SOM algorithm – which makes a learning update of the neuron weights after each record being read and matched – the Batch-SOM takes a "batch" of data (typically all records) and performs a "collected" update of the neuron weights after all records have been matched. This is much like "epoch" learning in supervised neural networks. The Batch-SOM is a more robust approach, since it mediates over a large number of learning steps. Most important, no learning rate is required (thus also no tuning!). Viscovery combines four enhancements to the plain Batch-SOM algorithm:

(i)   Most important, the "growing" of a SOM grid is used to speed up the training phase (see, for example, Kohonen, 1997, p. 141). This results in a speed-up factor of 5 to 10.

(ii)   A "quick match" algorithm is activated automatically in some of the learning schedules (very large numbers of records) which brings the learning time down further by a factor of 10.

(iii) Internally it uses a small (not really sensible, but necessary in special cases) momentum in the update process (as in supervised neural networks); in "Kohonen terms" this is the *Wegstein* factor.

(iv)   The neighborhood function is not a "cylinder" but a Gaussian.

Because of (i) and (ii) Viscovery may very well be the fastest implementation of the "original" SOM algorithm currently available, i.e. the resulting SOMs are practically the same as the results obtained from the latter. Much of this new approach is derived from the work of Teuvo Kohonen and his team.

Much like some other SOM tools, Viscovery creates a map in a two-dimensional hexagonal grid. Starting from numerical, multivariate data, the nodes on the grid gradually adapt to the intrinsic shape of the data distribution. Since the order on the grid reflects the neighborhood within the data, features of the data distribution can be read off from the emerging map on the grid.

In Viscovery, the trained SOM is systematically converted into visual information. The tool provides an extensive built-in capability for both pre-processing and post-processing as well as for the automatic colorcoding of the map and its components. The user gets all essential features for building and handling SOM applications and is moreover supported in advanced analysis tasks, such as

- searching for data clusters;
- retrieving numerical information and cluster statistics;
- monitoring of new data;
- evaluating dependencies between variables;
- investigating geometric properties of the data distribution.

Viscovery has been developed for professional users in business, industry, and science. It is particularly useful in the determination of dependencies between variables as well as in the analysis of high-dimensional cluster distributions. The tool supports the following important tasks for scientific and research purposes: dependency analysis, deviation detection, unsupervised clustering, non-linear regression, data association, pattern recognition, and animated monitoring.

A few real-world applications for which Viscovery has been used are

- market and customer profiling;
- customer scoring and behavior analysis;
- financial and economic modeling;
- medical applications;

- knowledge management and discovery in databases;
- industrial process optimization and quality control;
- scientific research.

Viscovery provides a powerful means to analyze complex data sets without prior knowledge of the basic SOM algorithm (although it helps if you read most of this book ...). The user is guided through the process by clear instructions; advanced users can tune the SOM creation parameters and make optimal use of the methods for data pre-processing. A comparison of the main features of Viscovery with those of the public domain software tools discussed earlier is shown in Table 13.1. An illustration of the use of Viscovery can be found in Chapter 15.

Additional sources for further research of SOM related topics on the web are provided in Box 2.

Box 2

In this book we discuss only one kind of SOM. There are several other kinds of SOM networks that have not been described here. They including among others:

DEC – Dynamically Expanding Context
LSM – Learning Subspace Method
ASSOM – Adaptive Subspace SOM
FASSOM – Feedback-controlled Adaptive Subspace SOM
Supervised SOM
LVQ-SOM – Linear Vector Quantization SOM
For more on-line information on these and other varieties of SOMs, see:

- The web page of The Neural Networks Research Center, Helsinki University of Technology, at

http://nucleus.hut.fi/nnrc/

- Akio Utsugi's web page on Bayesian SOMs at the National Institute of Bioscience and Human-Technology, Agency of Industrial Science and Technology, M.I.T.I., 1-1, Higashi, Tsukuba, Ibaraki, 305 Japan, at

http://www.aist.go.jp/NIBH/~b0616/Lab/index-e.html

Table 13.1 Comparison between commercial and public domain SOM software tools

	Viscovery SOMine	SOM_Pak	SOM Toolbox	NeNet
Operating system	Windows 95 Windows NT 4.0	UNIX MS DOS	requires MatLab Version 5 or higher	Windows 95 Windows NT
Data pre-processing	4 scaling options histogram modification variable transformation priority setting	none	variance-based plus others	variance-based range-based
SOM features • **Algorithm**	accelerated batch SOM	standard SOM	standard SOM batchmap	standard SOM
• **Map size**	unlimited	unlimited	unlimited	unlimited
• **Map topolopgy**	hexagonal	rectangular or hexagonal	rectangular or hexagonal	rectangular or hexagonal
• **Map initialization**	principal plane	linear or random	linear or random	linear or random
• **Training**	predefined schedules	any number of stages	any number of stages	any number of stages
• **Labeling**	automatic or by hand or drag & drop	automatic or by hand	automatic or by hand	automatic or by hand
• **Missing entries**	can be handled	can be handled	can be handled	can be handled
• **Speed**	very high	very high	moderate	high
• **Limits on inputs**	none	none (buffered loading)	none views	max 100 uni 100
• **Visualization**	cluster windows U-matrix component planes trajectories iso-contours	U-matrix component planes trajectories	U-matrix component planes trajectories hit histograms	U-matrix component planes trajectories hit histograms
• **Monitoring quality**	neighborhoods, paths monitoring value checks of neurons, frequency, quantization and curvature	quantization error Sammon weights fixed	quantization error Sammon weights component mask	easy inspection of neurons
Data postprocessing	summary statistics association/recall area selection, filtering	none	none	easy inspection of neurons
User interfaces	GUI user-friendly & interactive full OLE interface; MS Excel, text files SQL & DB2 coming	command line C programming text files	MatLab plus GUI MatLab language	GUI (conformation Windows 95) C++ text files
Installation	like Windows programs easy to install and uninstall	like PD-soft in UNIX	like toolboxes in MatLab	like Windows programs
Printing	standard Windows BMP and WMF option	EPS	any format supported by MatLab	standard Windows
Help	on-line help	manual	WWW doc on-line help	on-line help
Documentation	user's guide	program description	under development	under development

Appendix: Overview of Commercially Available Software Tools for Applying SOM

SAS Neural Network Application
(Company: SAS Institute Inc., email: software@sas.sas.com)

The SAS Neural Network Application trains a variety of neural nets and includes a graphical user interface, on-site training and customization. Features include multilayer perceptrons, radial basis functions, statistical versions of counterpropagation and learning vector quantization, a variety of built-in activation and error functions, multiple hidden layers, direct input-output connections, missing value handling, categorical variables, standardization of inputs and targets, and preliminary optimizations from random initial values to avoid local minima. Training is done by state-of-the-art numerical optimization algorithms instead of tedious backprop.

NeuralWorks
(Company: NeuralWare Inc., email: sales@neuralware.com)

Supports over 30 different nets: backprop, art-1, kohonen, modular neural network, general regression, fuzzy art-map, probabilistic nets, self-organizing map, lvq, boltmann, bsb, spr, etc. Extendable with optional package: ExplainNet, Flashcode (compiles net in .c code for runtime), user-defined i-o in c possible. ExplainNet (to eliminate extra inputs), pruning, savebest, graph.instruments like correlation, hinton diagrams, rms error graphs, etc.

MATLAB Neural Network Toolbox
(Company: The MathWorks Inc., email: info@mathworks.com)

The Neural Network Toolbox is a powerful collection of MATLAB functions for the design, training, and simulation of neural networks. It supports a wide range of network architectures with an unlimited number of processing elements and interconnections (up to operating system constraints). Supported architectures and training methods include: supervised training of feedforward networks using the perceptron learning rule, Widrow–Hoff rule, several variations on backpropagation (including the fast Levenberg–Marquardt algorithm), and radial basis networks; supervised training of recurrent Elman networks; unsupervised training of associative networks including competitive and feature map layers; Kohonen networks, self-organizing maps, and learning vector quantization. The Neural Network Toolbox contains a textbook-quality Users' Guide, uses tutorials, reference materials and sample applications with code examples to explain the design and use of each network architecture and paradigm. The Toolbox is delivered as MATLAB M-files, enabling users to see the algorithms and implementations, as well as to make changes or create new functions to address a specific application.

NeuroShell2/NeuroWindows
(Company: Ward Systems Group, Inc., email: WardSystems@msn.com)
NeuroShell 2 combines powerful neural network architectures, a Windows icon-driven user interface, and sophisticated utilities for MS-Windows machines. Internal format is spreadsheet, and users can specify that NeuroShell 2 use their own spreadsheet when editing. Includes both beginner's and advanced systems, a runtime capability, and a choice of 15 backpropagation, Kohonen, PNN and GRNN architectures. Includes rules, symbol translate, graphics, file import/export modules (including MetaStock from Equis International) and NET-PERFECT to prevent overtraining.

NeuroWindows is a programmer's tool in a Dynamic Link Library (DLL) that can create as many as 128 interactive nets in an application, each with 32 slabs in a single network, and 32K neurons in a slab. Includes backpropagation, Kohonen, PNN, and GRNN paradigms. NeuroWindows can mix supervised and unsupervised nets. The DLL may be called from Visual Basic, Visual C, Access Basic, C, Pascal, and VBA/Excel 5.

NeuroSolutions v3.0
(Company: NeuroDimension, Inc., email: info@nd.com)

NeuroSolutions is a graphical neural network simulation tool. Because of its object-oriented design, NeuroSolutions provides the flexibility needed to construct a wide range of learning paradigms and network topologies. Its GUI and extensive probing ability streamline the experimentation process by providing real-time analysis of the network during learning.

Topologies:

- multilayer perceptrons (MLPs);
- generalized Feedforward networks;
- modular networks;
- Jordan–Elman networks;
- self-organizing feature map (SOFM) networks;
- radial basis function (RBF) networks;
- time delay neural networks (TDNN);
- time-lag recurrent networks (TLRN);
- user-defined network topologies.

NeuroLab, A Neural Network Library
(Company: Mikuni Berkeley R&D Corporation,
e-mail: neurolab-info@mikuni.com)

NeuroLab is a block-diagram-based neural network library for Extend simulation software (developed by Imagine That Inc.). The library aids the understanding, designing and simulating of neural network systems. The library consists of more than 70 functional blocks for artificial neural network implementation and many example models in several professional fields. The package provides icon-based functional blocks for easy implementation of simulation models. Users click, drag and connect blocks to construct a neural network and can specify network parameters – such as back propagation methods, learning rates, initial weights, and biases – in the dialog boxes of the functional blocks. Users can modify blocks with the Extend model-simulation scripting language, ModL, and can include compiled program modules written in other languages using XCMD and XFCN (external command and external function) interfaces and DLL (dynamic linking library) for Windows. The package provides many kinds of output blocks to monitor neural network status in real time using color displays and animation and includes special blocks for control application fields. Educational blocks are also included for people who are just beginning to learn about neural networks and their applications. The library features various types of neural networks –including Hopfield, competitive, recurrent, Boltzmann machine, single/multilayer feedforward, perceptron, context, feature map, and counter-propagation – and has several backpropagation options: momentum and normalized methods, adaptive learning rate, and accumulated learning.

havFmNet++
(Company: hav.Software, email: hav@neosoft.com)

havFmNet++ is a C++ class library that implements self-organizing feature map nets. Map-layers may be from one to any dimension. havFmNet++ may be used for both stand-alone and embedded network training and consultation applications. A simple layer-based API, along with no restrictions on layer-size or number of layers, makes it easy to build single-layer nets or much more complex multiple-layer topologies. havFmNet++ is fully compatible with havBpNet++ which may be used for pre- and post-processing. Supports all standard network parameters (learning-rate, momentum, neighborhood, conscience, batch, etc.). Uses On-Center-Off-Surround training controlled by a sombrero form of Kohonen's algorithm. Updates are controllable by three neighborhood-related parameters: neighborhood-size, block-size and neighborhood-coefficient cutoff. Also included is a special scaling utility for data with large dynamic range. Several data-handling classes are also included. These classes, while not required, may be used to provide convenient containers for training and consultation data. They also provide several normalization/de-normalization methods.

Neural Connection
(Company: Neural Connection, email: sales@spss.com)

Neural Connection is a graphical neural network tool which uses an icon-based workspace for building models for prediction, classification, time-series forecasting and data segmentation. It includes extensive data management capabilities so your data preparation is easily done right within Neural Connection. Several output tools give you the ability to explore your models thoroughly so you understand your results. Modeling and forecasting tools include: multi-layer perceptron, radial basis function, and Kohonen network. Statistical analysis tools include multiple linear regression, closest class means classifier, and principal component analysis

Trajan 2.0 Neural Network Simulator
(Company: Trajan Software Ltd., email: andrew@trajan-software.demon.co.uk)

Trajan 2.1 Professional is a Windows-based neural network tool that includes support for a wide range of neural network types, training algorithms, and graphical and statistical feedback on neural network performance. Features include:

1. Full 32-bit power. Trajan 2.1 is available in a 32-bit version for use on Windows 95 and Windows NT platforms, supporting virtually unlimited network sizes (available memory is a constraint). A 16-bit version (network size limited to 8192 units per layer) is also available for use on Windows 3.1.
2. Network architectures. Includes support for multilayer perceptrons, Kohonen networks, radial basis functions, linear models, probabilistic and generalized regression neural networks. Training algorithms include the very fast, modern Levenburg–Marquardt and conjugate gradient descent algorithms, in addition to backpropagation (with time-dependent learning rate and momentum, shuffling and additive noise), quick propagation and delta-bar-delta for multilayer perceptrons; K-means, K-nearest neighbor and pseudo-inverse techniques for radial basis function networks, principal components analysis and specialized algorithms for automatic network design and neuro-genetic input selection. Error plotting, automatic cross verification and a variety of   stopping conditions are also included.
3. Custom architectures. Trajan allows you to select special activation functions and error functions; for example, to use softmax and cross-entropy for probability estimation, or city-block error function for reduced outlier-sensitivity. There are also facilities to "splice" networks together and to delete layers from networks, allowing you to rapidly create pre- and post-processing networks, including autoassociative dimensionality reduction networks.
4. Simple user interface. Trajan's carefully-designed interface gives you access to large amounts of information using graphs, bar charts and datasheets. Trajan automatically calculates overall statistics on the performance of networks in both classification and regression. Virtually all information can be transferred via the clipboard to other Windows applications such as spreadsheets.
5. Pre- and post-processing. Trajan 2.1 supports a range of pre- and post-processing options, including minimax scaling, winner-takes-all, unit-sum and unit-length vector. Trajan also assigns classifications based on user-specified accept and reject thresholds.
6. Embedded use. The Trajan Dynamic Link Library gives full programmatic access to Trajan's facilities, including network creation, editing and training. Trajan 2.1 come complete with sample applications written in "C" and Visual Basic. There is also a demonstration version of the software available; please download this to check whether Trajan 2.1 fulfils your needs.

14 Tips for Processing and Color-coding of Self-Organizing Maps

Samuel Kaski and Teuvo Kohonen

Abstract

The most important aspects for creating good SOM maps are the selection of the size and shape of the SOM, the scaling of the input variables, the selection of the neighborhood function and the learning rate, and the initialization of the model vectors. In addition, we discuss automatic procedures for color-coding of SOM maps. The data entries corresponding to, for example, different countries can be automatically colored so that similar data attain a similar color.

14.1 The SOM Array

In order to make a good visual display, the arrangement of the neurons in the array ought to be hexagonal. The array should be oblong, because the distribution of the data samples is usually oblong, and the net of the model vectors should be fitted to it. Square arrays usually do not orient themselves well enough with the distribution.

The number of neurons in the array is a rather delicate figure that depends on the application and the amount of available data. For instance, if you have only one set of indicators per entry then the number of neurons might be selected as somewhat smaller than the number of entries. If, on the other hand, you have sample vectors that are stochastic variables that fall into more or less fuzzy clusters and you are interested in showing the cluster structure on the SOM, then the number of neurons would be better as a multiple of the number of clusters.

Sometimes the number of available samples is limited due to the high costs or difficulties in acquiring them. In such a case the statistical accuracy of representation may become a problem; you cannot fit many nodes to a single sample. If you are studying clustering by the SOM, the number of neurons should not be higher than, say, a fraction of the order of 10% of the total number of samples. So there are many different cases, and it is essential to understand the nature of the problem and the data before trying to display it with SOM.

14.2 Scaling the Input Variables

In some cases it may perhaps be surprising that you have different indicators with different weights, but nonetheless you may have to equalize the scales of the different indicators. Equal scaling, scaling of the variance of each indicator to unity, is in general a good first choice in preliminary clustering tests, and effective even in the final maps. However, sometimes, especially when concentrating on particular indicators, it may turn out to be useful to try different scales for various indicators. Variable scaling is closely connected with the selection of the most important input variables, as mentioned elsewhere in this book.

14.3 Initialization of the Algorithm

A very special property of the SOM algorithm is that the initial values for the model vectors can be selected as arbitrary random vectors. The algorithm is able to scale them properly and sort them into a smooth order in the long run. Selection of random initial values, however, is not at all the most reasonable or effective way to proceed in practice. The self-organizing process works orders of magnitude faster, and the final results are much more stable, if you set a preliminary rough order for the model vector values before starting the algorithm. For instance, the SOM_PAK software has a preprogrammed provision for automatic determination of the two largest principal components of the data samples, upon which a regular grid of points can be spanned along the hyperplane defined by the sample data. These values should be taken for the initial values of the model vector grid; this allows much narrower neighborhood functions and a smaller learning-rate factor.

14.4 Selection of the Neighborhood Function and Learning Rate

In extensive tests it has turned out that the SOM algorithm tolerates very different choices for the neighborhood function and the learning-rate factor. However, what seems to be most essential is that

- the neighborhood function should be wider in the beginning of the learning process, and the width should decrease with time so that at the end of the process only the immediate neighbors of the winner are updated;
- there should be enough training steps in the self-organizing process.

While the Gaussian neighborhood function introduced in Chapter 11 may be a little difficult to use, it is quite possible to resort to the following simpler choice for neighborhood function that also works fairly well, especially if the model vectors are initialized as recommended above.

The simpler SOM equations are

$$\mathbf{m}_i(t+1) = \mathbf{m}_i(t) + \alpha(t)[\mathbf{x}(t) - \mathbf{m}_i(t)] \text{ if } i \text{ belongs to } N_c(t),$$

$$\mathbf{m}_i(t+1) = \mathbf{m}_i(t) \text{ if } i \text{ does not belong to } N_c(t).$$

The exact choice of $N_c(t)$ and $\alpha(t)$ is not critical, if you start with the preordered initial values of the $\mathbf{m}_i$. You can then let the radius of N_c decrease linearly (i.e., in discrete steps) from, say, 3 to 1, and at the same time decrease the learning-rate from, say, 0.02 to zero.

The number of steps of the algorithm should be at least 100 times the number of nodes (neurons) in the SOM to achieve best possible results, preferably even more.

In order to achieve even better convergence, theoretical considerations favor a law of the type $\alpha(t) = 0.02\,T\,/(T + t)$, where T is a parameter that may have the value of 100 times the number of neurons or more.

The above recommendations for numerical choices of $N_c(t)$ and $\alpha(t)$ do not hold if the model vectors are initialized with random values. In this case the initial radius of N_c should be half of the diameter of the net, and the learning rate should be close to unity in the beginning. These choices have been used in many experiments, however, we do not recommend random initialization in practice! We recommend regular initial values, because there is no need to demonstrate "self-organization from scratch" every time.

14.5 Automatic Color-coding of Self-Organizing Maps

To make the display generated by a self-organizing map more intuitive different data clusters, i.e., types of data, can be encoded with different colors. The procedure described below is an automatic procedure, after which the user can still change the coloring interactively.

If we want to color a SOM map according to the clusters it represents, we could simply color each cluster with a different color. It is often not possible, however, to extract clear-cut clusters because they may be overlapping, and it is even possible that there are no clusters in the data. Therefore, the data should be colored so that the hues change slowly with the density of the data. Clusters which are relatively more homogeneous will be more uniformly colored, while the areas of the map which do not correspond to clusters will be more heterogeneously colored.

Our coloring procedure consists of three successive steps: (i) selection of *color centers,* suitable locations of the map that will be allocated known colors; (ii) selection of suitable colors for the color centers; (iii) coloring of the rest of the map.

14.5.1 Selection of the Color Centers

We used a SOM to select the color centers: each node of a circular (one-dimensional) SOM can be considered to represent one color center, and when the SOM is trained in a proper manner the color centers end up predominantly in the clustered areas.

The circular SOM chain can be taught using the ordinary SOM algorithm but with specially generated input data. The motivation for how the input data is created can be seen in the display in Figure 14.1, where the clustered areas are denoted by light shades. We choose two-dimensional data, in which the first

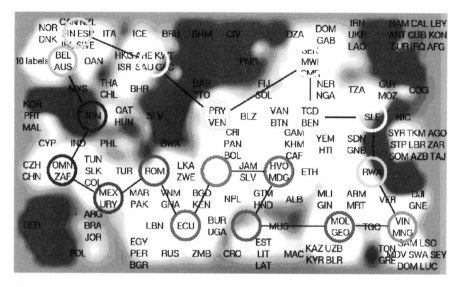

Figure 14.1. Structured map display of the poverty types in the world. The data consisted of 39 indicators of poverty selected from the World Development Indicators published by the World Bank in 1992. The automatically chosen color centers have been encircled with the colors they will receive. The gray scale in the background display reflects the clustering tendency of the data. Light areas denote clusters and dark areas rapid changes in the poverty type.

component corresponds to the horizontal axis of the (two-dimensional) SOM display and the second component corresponds to the vertical axis. More data is chosen from the white areas and less data from the darker regions – the data will be chosen as if the shades in Figure 14.1 were used to depict the probability density function of the data. It is known that the model vectors of the SOM reflect the distribution of the data; therefore they will predominantly end up on the clustered (white) areas. The actual procedure of creating the data involves some additional technical details which we shall omit here.

When a SOM chain consisting of 18 nodes is used to place the color centers they will be located as shown in Figure 14.1.

14.5.2 Selection of Colors for the Color Centers

If each color center were placed manually onto a separate clustered area it would be justified to simply assign a different color to each of them. Every possible hue can be expressed as an angle of a color circle, in which different hues are located at different angles. The hues of the manually chosen color centers could then be chosen to be equally spaced on the color circle. Nearby color centers should be assigned similar hues if possible.

When the color centers are placed automatically on the SOM display, several centers may end up in the same large clustered area. For example, the large white area around the center of Figure 14.1 contains a large number of color centers.

Therefore the hues of the color centers cannot be chosen equally spaced or otherwise the color would change very rapidly within the clustered areas.

Suppose that the hues of the cluster centers were chosen to be mutually relatively more similar in the clustered areas. Then each cluster would receive a distinctive hue that would change slowly towards the surroundings of the cluster. In between the clusters the color would change rapidly. That kind of a coloring is precisely what we are trying to construct, and it can be achieved by using a chain-like SOM to place the color centers. The circular SOM chain can be regarded as a "hue circle" that is being fitted to the display. The nodes of the circular SOM will be assigned hues with a spacing that is approximately inversely proportional to the "degree of clustering" or density of the input data, whereby the color centers that lie within the same cluster will be assigned similar hues.

In order to understand the technical details of the procedure it may be useful to note that each color center is attached to one node on the two-dimensional map that forms the display groundwork in, for instance, Figure 14.1. The "clustering tendency" around each color center can therefore be approximated based on the distances of the neighboring model vectors of the two-dimensional map, since the density of the model vectors reflects the density of the data.

Specifically, the spacing of the hues will be determined by the distances of the model vectors that correspond to the color centers on the two-dimensional SOM. The distance between the centers is computed along a path that starts from one of the color centers, passes at each step to a model vector of a neighboring node, and finally ends up in the other color center. From all such paths we choose the shortest one. This procedure may seem a bit complicated but it is needed to ensure that the rest of the map will be colored consistently.

The circles around the color centers in Figure 14.1 have been assigned hues according to the procedure described above. In the top left corner, for example, there are two similarly colored color centers. In the large clustered area around the center of the display there are several color centers. That clustered area is so large that the hue changes clearly even within the cluster. It is in fact quite reasonable that a large cluster is not colored homogeneously; if the whole map consisted of a single large cluster it would be reasonable to display the structures within this cluster instead of assigning the same hue to the whole map.

14.5.3 Coloring of the Rest of the Map

Now that the cluster centers have been chosen and colored it is time to color the rest of the map display. Each location is assigned a color that reflects its similarity with the color centers. Each location will receive a mixture of the colors of the color centers. Each mixing coefficient will be larger the more similar the location is to the color center.

Specifically, each mixing coefficient will be a function of the distance between the model vector at the map location to be colored and one of the color centers. The distance will again be computed along the shortest path that goes at each step from one model vector to a neighboring one. The mixing coefficient will be a Gaussian function of the distance. The inverse of the distance would be another

Table 14.1. Key to the country symbols in Figures 14.1 and 14.2

Symbol	Country	Symbol	Country	Symbol	Country
AFG	Afghanistan	GRC	Greece	NOR	Norway
AGO	Angola	GTM	Guatemala	NPL	Nepal
ALB	Albania	HKG	Hong Kong	NZL	New Zealand
ARE	United Arab Emirates	HND	Honduras	OAN	Taiwan China
ARG	Argentina	HTI	Haiti	OMN	Oman
AUS	Australia	HUN	Hungary	PAK	Pakistan
AUT	Austria	HVO	Burkina Faso	PAN	Panama
BDI	Burundi	IDN	Indonesia	PER	Peru
BEL	Belgium	IND	India	PHL	Philippines
BEN	Benin	IRL	Ireland	PNG	Papua New Guinea
BGD	Bangladesh	IRN	Iran Islamic Rep.	POL	Poland
BGR	Bulgaria	IRQ	Iraq	PRT	Portugal
BOL	Bolivia	ISR	Israel	PRY	Paraguay
BRA	Brazil	ITA	Italy	ROM	Romania
BTN	Bhutan	JAM	Jamaica	RWA	Rwanda
BUR	Myanmar	JOR	Jordan	SAU	Saudi Arabia
BWA	Botswana	JPN	Japan	SDN	Sudan
CAF	Central African Rep.	KEN	Kenya	SEN	Senegal
CAN	Canada	KHM	Cambodia	SGP	Singapore
CHE	Switzerland	KOR	Korea Rep.	SLE	Sierra Leone
CHL	Chile	KWT	Kuwait	SLV	El Salvador
CHN	China	LAO	Lao PDR	SOM	Somalia
CIV	Cote d'Ivoire	LBN	Lebanon	SWE	Sweden
CMR	Cameroon	LBR	Liberia	SYR	Syrian Arab Rep.
COG	Congo	LBY	Libya	TCD	Chad
COL	Colombia	LKA	Sri Lanka	TGO	Togo
CRI	Costa Rica	LSO	Lesotho	THA	Thailand
CSK	Czechoslovakia	MAR	Morocco	TTO	Trinidad and Tobago
DEU	Germany	MDG	Madagascar	TUN	Tunisia
DNK	Denmark	MEX	Mexico	TUR	Turkey
DOM	Dominican Rep.	MLI	Mali	TZA	Tanzania
DZA	Algeria	MNG	Mongolia	UGA	Uganda
ECU	Ecuador	MOZ	Mozambique	URY	Uruguay
EGY	Egypt Arab Rep.	MRT	Mauritania	USA	United States
ESP	Spain	MUS	Mauritius	VEN	Venezuela
ETH	Ethiopia	MWI	Malawi	VNM	Vietnam
FIN	Finland	MYS	Malaysia	YEM	Yemen Rep.
FRA	France	NAM	Namibia	YUG	Yugoslavia
GAB	Gabon	NER	Niger	ZAF	South Africa
GBR	United Kingdom	NGA	Nigeria	ZAR	Zaire
GHA	Ghana	NIC	Nicaragua	ZMB	Zambia
GIN	Guinea	NLD	Netherlands	ZWE	Zimbabwe

alternative but we use the Gaussian function to avoid extensive "leaking" of the color from the color centers to distant positions on the map.

There is still one technical detail left that is probably worth mentioning. Each mixing coefficient is weighted by the inverse of the density of the other color centers around the color center, estimated based on the distances between the neighboring color centers. If such a weighting were not used the color of a large clustered area containing several color centers would spread more strongly to the surroundings than the color of a smaller cluster.

It may be useful to note, to aid in the interpretation of the coloring, that the resolution of the coloring is at the highest around the clustered areas. The resolution is governed by the density of the color centers, and the color centers will be placed around the clusters in order to guarantee that the clusters will receive pure, bright colors. Therefore, locations of the map that are far away from the clustered areas are likely to receive the color of the closest clustered area. This seems to be a viable strategy although the resolution of the coloring may become low far from the clustered areas.

Some examples of color-coding of SOM maps are shown in Figures 14.2 and 14.3. These maps are based on 39 indicators of poverty selected from the World Bank Development Indicators published in 1992. The legend for the country symbols in these maps can be found in Table 14.1.

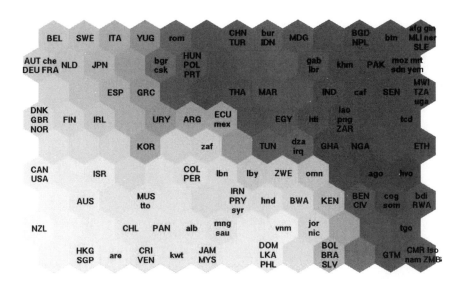

Figure 14.2. A SOM of world poverty, based on 39 indicators selected from the World Development Indicators published by the World Bank. In this map different colored cluster areas represent different poverty types. The color changes gradually among the clusters; similarity of poverty types is reflected in the similarity of colors. *(This figure can be seen in the Colour Plate Section.)*

Figure 14.3. A world map based on poverty types obtained via a self-organizing map applied to 39 indicators selected from the World Development Indicators. (*This figure can be seen in the Colour Plate Section.*)

15 Best Practices in Data Mining using Self-Organizing Maps

Guido Deboeck

Abstract

This chapter summarizes best practices in data mining and visual data explorations through clustering of multi-dimensional data in finance, economics and marketing. The best practices outlined in this chapter are based on (i) the lessons learned from all the chapters in this book, (ii) lessons learned from other papers not included here, (iii) the expertise of people who have several years of hands-on experience in applying neural networks in finance and economics. From the applications presented in this book we derived a process for data analysis, clustering, visualization, and evaluation in finance, economics and marketing. This chapter outlines this process and illustrates it by applying it to country credit risks analysis.

15.1 Main Steps in using Self-Organizing Maps

The financial, economic and marketing applications of self-organizing maps outlined in this book show that there are no specific procedures or optimal methods for applying SOM that are valid for all applications. Similar to the design of other neural network models, the creation of a self-organizing map is still an art more than a science.

Chapter 14 contained tips for the selection of a SOM array; the scaling of the input variables; the initialization of the algorithm; the selection of the neighborhood size and the learning rate; and the interpretation and color-coding of the map. These relate to what may be called the "engineering" aspects of SOM. The tips provided in Chapter 14 are very important but not sufficient for the entire data analysis process covering clustering, visualization, interpretation and evaluation of results. In this chapter we expand on these and outline "best practices" for data explorations in finance with SOM.

The best practices outlined in this chapter are based on (i) the lessons learned from all the chapters in this book, (ii) lessons learned from other papers not included here, (iii) the expertise of people who have several years of hands-on experience in applying neural networks in finance and economics.

From the applications presented in this book we can derive the main steps of a process for data analysis using SOM. These are presented in Box 1 and are discussed in detail in this chapter.

Box 1: Main steps in clustering and visualization of data using SOM

Step 1. Define the purpose of the analysis.

Step 2. Select the data source and quality.

Step 3. Select the data scope and variables; consider if combinations of variables could describe your system better.

Step 4. Decide how each of the variables will be preprocessed.

Step 5. Use relevant sample data that are representative for your system.

Step 6. Select the clustering and visualization method(s); consider the use of hybrid methods.

Step 7. Determine the desired display size, map ratio, and the required degree of detail.

Step 8. Tune the map for optimal clustering and visualization.

Step 9. Interpret the results, check the values of individual nodes and clusters.

Step 10. Define or paste appropriate map labels.

Step 11. Produce a summary of the results highlighting the differences between clusters.

Step 12. Document and evaluate the results.

15.1.1 Define the Purpose of the Analysis

Without proper definition of the goals and objectives for the design of a neural network model, supervised or unsupervised, it will be difficult to assess the effectiveness of the outcome. Neural net models can be designed for many different objectives. As shown in several applications in this book the main objectives of creating self-organizing map can be for:

(i)   classification, clustering, and/or data reduction;
(ii)   visualization of the data;
(iii)   decision-support;
(iv)   hypothesis testing;
(v)   monitoring system performance;
(vi)   lookup of (missing) values;
(vii)   forecasting.

If clustering and visualization of the data are the main objectives, various alternative visualization and clustering methods should be considered. Several traditional statistical methods for clustering and data visualization exist (see Introduction). Combining traditional statistical methods with SOM may generate better results than the use of SOM by itself. It may also be useful to determine a priori how much data reduction is desired.

If decision-support is the main objective then it is essential to define precisely what decisions need to be supported, what is the scope of these decisions, and what is their time frame. For example, mapping of information on mutual funds to support the selection of managers to be included in a portfolio (see Chapter 3) has a different scope than using SOM for mapping compliance to an accord for the integration of countries into the European monetary union (as discussed in Chapter 1).

If hypothesis testing is the main objective one needs to define a priori what hypotheses will be tested and what will be the standard for acceptance or rejection. For example, when applying SOM to the IFC data on banking institutions in emerging markets (see Chapter 6), the hypothesis we tested was: is there a significant difference between the various banking institutions in various emerging markets around the world?

If monitoring systems performance is the objective the goals of the monitoring process need to be defined, e.g. monitoring for quality purposes, fault detection, standard compliance.

If forecasting is the objective, it is important to spell out what is the forecasting window, the desired accuracy, and how the performance will be evaluated. For example, in using SOM to predict stock or commodity prices, what time window are the predictions for (10 days, a month, 6 months); should the predictions be accurate in terms of the level or just the direction (up or down); will the price predictions be evaluated in terms of the percentage of correct predictions or in terms of the cumulative profit or loss achieved by implementing the predictions in a given period?

15.1.2 Select the Data Source and Data Quality

From several of the applications outlined in this book we learned the importance of using high-quality data. It is important that the data comes from one or more reputable sources. Good sources of high-quality financial and economic data are:

- national and international agencies (e.g. government statistical offices, the United Nations, specialized agencies of the UN, World Bank, IMF, and the like);
- well-established information services (e.g. Bloomberg, Reuters, Telerate, Knight-Ridder, Standard & Poor);
- database providers (e.g. Value Line, Morningstar, DRI, Moody's, American OnLine, CompuServe, etc.).

Data that is freely available on the web may or may not be of high quality. It is therefore advisable to be skeptical about what is freely offered on the web. In general one should choose data sources that are well established and recognized for providing high-quality data.

15.1.3 Select the Data Scope and Variables

To select or define the data scope in relation to the objectives of the study is important for any kind of data analysis. Competitive learning and the power of the SOM algorithm may cause laziness and tempt some to "throw the kitchen sink in", i.e. use all the available data on a particular subject rather than a selective set that is relevant to the objectives of the study.

Furthermore, it is important to have an understanding of the data, to use domain expertise, or to collaborate with those that have such expertise. For example, when investigating structures in investment data, credit risk data, or poverty data, proper SOM analyses cannot be done without domain knowledge.

One should also be very careful in the selection of the appropriate indicators. For example, poverty in developed countries is not necessarily the same or measured with the same indicators as poverty in developing countries; the opposite of poverty is not necessarily welfare. What is not in the data can also not be extracted.

Once the data scope has been properly defined, some important tips to remember in selecting variables to be included in the analysis are:

- do not get wed to your data, learn to discriminate, discard and delete;
- select only those variables that are meaningful in relation to the objectives;
- select the variables that are most likely to influence the results;
- consider using combinations of variables, such as ratios, time-invariants, etc.;
- use domain expertise or involve in the analysis people who have domain expertise;
- do not assume that the data is normally distributed;
- adding of one or more irrelevant variables can dramatically interfere with the cluster recovery;
- omission of one or more important variables may also affect the results.

15.1.4 Decide How Each of the Variables will be Preprocessed

Preprocessing of data is important in any data analysis, in particular in neural network design. Chapter 2 in *Trading on the Edge* (Deboeck, 1994) outlines several principles that apply to preprocessing of data for both supervised and unsupervised neural networks. In the latter case preprocessing may more specifically involve data standardization, data transformations, and setting of priority.

The main reason for *data standardization* is to scale all data to the same level. Often the data range of each variable varies from column to column. If no preprocessing is applied this may influence the clustering and the ultimate shape of a SOM map. There are many ways in which data can be standardized. The most common (as we have learned from the applications in this book) are to standardize all data based on the standard deviation. Other methods are to standardize on the basis of the range, e.g. $z = [x - \min(x)]/[\max(x) - \min(x)]$. Some studies have shown that standardizing the data based on the range can be superior in certain cases, in particular if the variance is much smaller than the range.

Data transformations can be applied to any or all variables to influence the importance and/or influence of each variable on the final outcome. Transformations may also be used to "equalize" the histograms. Two typical data transformations are *logarithmic* and *sigmoid*. The former squeezes the scale for large values, the latter takes care of outliers. Applying data transformations redefines the internal representation of each variable; they should be applied with caution.

Setting the priority of a variable to a value greater or lower than one has the same effect as changing the standardization explicitly. By giving a priority to a variable you can provide a weighting of the variables in the mapping process. For example, if in the selection of investment managers, the "launching date" of a mutual fund is considered less important, then this variable can be given a low priority.

15.1.5 Use Relevant Sample Data that are Representative for your System

Training a neural network on a set of sample data will yield better results when using prototype input vectors. By selecting "good", i.e. representative, input vectors for the map training we reduce noise and can obtain a sharper map. This map then can be used for testing on all the remainder input data sets. Furthermore, depending on the applications, the use of input vectors that represent outliers may be of crucial importance for training a SOM. Provided these vectors are prototypes, outliers provide contrasts and can sharpen the differences between clusters. However, this can be to the cost of sensibility for the other parts of the map. If outliers are not representative, they should of course be eliminated.

In training SOM maps we also learned that training in two stages is often preferred. Usually, the first training cycle is used to obtain a rough structure and the second cycle for capturing the finer differences between the input vectors. This dual training cycle can be implemented by choosing different learning rates and neighborhood sizes for each cycle. In the first cycle the learning rate will often start from 0.5 and be decreased over time. In the second training cycle the learning rate may be started at a much lower value, e.g. 0.02, and be decreased over time. The neighborhood size usually starts from half the diameter of the net, in order to shrink to zero. Newer SOM implementations require no more learning rates at all by "batching" all learning increments for several individual updates.

15.1.6 Select the Clustering and Visualization Method(s); Consider the use of Hybrid Methods

In this book we have focused on SOM. Several chapters have, however, demonstrated that by combining SOM with other methods better results can be obtained. Marina Resta showed in Chapter 7 that a hybrid system of SOM and genetic algorithms can improve the performance of trading models. Eero Carlson and Anna Tulkki talked in Chapters 8 and 9 about combining SOM with GIS.

In financial, economic and marketing applications combining SOM with other statistical methods is or may become common practice. Often in finance and economics it is important to undertake post-processing of SOM results. As we learned, a SOM map by itself provides a topological representation of the data. This representation still needs to be translated into operational or actionable outcomes. Financial analysts, economists and certainly marketing professionals will want to know what are the main features of the clusters represented on a SOM map, how they differ from each other, and how to use the newly found structures or patterns for forecasting or decision-support. Thus a SOM map by itself cannot be a final outcome.

15.1.7 Determine the Desired Display Size and Shape, and the Required Degree of Detail

Bigger maps produce more detail; input vectors are spread out on a larger number of nodes. Smaller maps can contain bigger clusters or more input vectors can cluster on a smaller set of nodes. Which is better? This will depend on the application and the usage of the map. Smaller is not necessarily better. More detail may be desirable in some cases. In general, smaller numbers of nodes stand for higher generalization, and this may also be useful if the data contains more noise. Higher numbers of nodes normally yield better map images but must not be over-interpreted in later use.

The key in determining the size of the map is how the map will be used. A simple analogue would be to compare the use of a country atlas with that of highway or street maps: at home I keep a global atlas for easy reference and lookup; however, when traveling I never use an atlas, not even a small one; highway and city maps are more useful for traveling. If using SOM for lookup of information on large data sets, a larger SOM map may be more desirable; however, when using SOM to select investment opportunities or investment managers, a smaller map that clusters managers and investment opportunities in five to seven categories may be more optimal.

The most frequently used methods to define the shape of the map are either to take the ratio of the principal plane or that of the Sammon map. Some tools provide an option to determine the map shape automatically.

15.1.8 Tune the Map for Optimal Clustering and Visualization

Once a SOM has been trained you can inspect the map by looking at the number of nodes that contain input vectors, the mean values of the nodes and clusters, the number of clusters that were created, and the number of matching input vectors for each cluster. Fine-tuning a map can be done by increasing or reducing the cluster threshold and/or the minimum cluster size (more on this in Section 15.2). A larger cluster threshold or higher minimum cluster size will reduce the number of clusters, it will increase the coarse-ness of the clustering. Lowering the cluster threshold will show more details of the map. This will be illustrated in Section 15.2.

15.1.9 Interpret the Results, Check Values of Individual Nodes and Clusters

Once a topological representation of the data is created, it is important to check the validity of the map. This can be done in several ways. Again, domain expertise will be a key ingredient. A simple check may consist in printing a list of the input vectors sorted by node or cluster of the map. Another one may be to calculate some simple summary statistics on each cluster. The tables in Chapters 3 and 6 were examples of this. Depending on which software tool is used, the mean values of the clusters may be even displayed on the screen. In this case the user can interactively check each cluster and judge whether the summary values make sense. Comparisons of values among nodes and clusters will then allow the user to decide on how much more detailed the map needs to be, which data transformation could be needed, how to fine-tune the priority of some components, or what the generalization capability of the map eventually may be. In other words, an interactive capability to check the values for nodes and clusters is important in order to allow the process to be dynamic and to incorporate the user's domain expertise and knowledge about the data.

15.1.10 Define or Paste Appropriate Map Labels

The importance and difficulty of defining appropriate labels has been discussed several times in this book. When using SOM to classify countries, states or cities, or when using SOM to cluster investment opportunities, companies or banks, the labels to be used are obvious: each input vector can be extended with an appropriate or abbreviated name of the country, state, city, security, company or bank it represents. When using SOM for process control labeling may be restricted to a few input vectors, picking on those that represent failures or idle states. When using SOM to classify wines or whiskeys, multiple labels may be necessary to identify the country, region, vineyard or distillery. In sum, flexibility in automatic labeling of nodes or clusters from the input data vectors is of crucial importance. This automatic labeling capability is of particular importance for finance, economic and marketing applications.

15.1.11 Produce a Summary of the Map Results that Highlights the Differences between Clusters

The production of summary statistics may be automatic or manual depending on which software tool is used for SOM. Newer software packages have built-in capabilities for automatic calculation of summary statistics. This has clear advantages over software tools that do not provide any post-processing capability. In finance, economics and marketing, post-processing of SOM results, information extraction of value added, and how SOM results can be used is very important. A post-processing capability that allows to create summary statistics for each node and each cluster, showing at the minimum the mean, standard deviation, minimum and maximum values, and the sum of the input vectors, is a great advantage. An example of such a post-processing capability is shown in Section 15.2.

15.1.12 Document and Evaluate SOM Results

For SOM to be useful in finance, economics and marketing, it is essential to demonstrate its value added. "Look Mom, what a nice picture I made" will just not fly in boardrooms, management meetings, or strategic marketing sessions. When we applied supervised neural networks to create financial models, we measured the value added of those models by measuring the performance (return), the risks, and the portfolio turnover of the models; we compared results with those of benchmarks (e.g. performance of human traders, or models based on more traditional methods).

Return is usually compared with risk to obtain the risk-adjusted return. This risk-adjusted return can be compared with a benchmark (e.g. the risk-adjusted return of the Standard and Poor 500). By adding portfolio turnover, one can take into account the costs of trading. The higher the turnover, the higher the transaction costs. The tradeoff between risk-adjusted return and costs then provides a measure of effectiveness of trading models.

If we apply the same approach to clustering and data visualization, i.e. find a measure of "return", "risks" and "turnover", we would be able to compare those measures with similar measures of benchmarks. In Box 2 we elaborate on how these measures could be applied to clustering and visualization. A methodology for the evaluation of SOM results is essential and important for more advanced applications in finance, economics and marketing.

Box 2: A methodology for the evaluation of SOM in finance and economics

(i) The outcome of clustering is a *number of data clusters*: the maximum is presumably equal to the number of observations in the data if the size of the map is very large; the minimum is one if all the training data is very similar or zero if there is no clustering tendency in the data. It is safe to assume that the desired outcome will probably be closer to the minimum if data reduction is one of the objectives. The optimal number of clusters will, however, depend on the data, the objectives of the study and how the map will be used. The number of clusters defined by a map may be considered as the "return" from clustering.

(ii) Another outcome of clustering is the degree of differences between the clusters. There could be a small or a large number of clusters, and very similar or very different clusters. This fuzzy outcome provides a scope for the *quality of clustering*. Suppose that the quality of clustering is measured on a scale from 0 to 100, then the cluster quality is low if there is little clustering tendency, a small number of clusters or small differences between clusters. The clustering quality is high if the clustering produces a lot of clusters, or clusters that are very different (that depending on the objectives of the study can have multiple usage). Low clustering quality indicates that there is not much differentiation between the clusters obtained via SOM; high clustering quality may indicate that a lot of differentiation is found. Of course, this is only one way of defining quality of clustering. Just like "risk" in investment terms, there may be many other ways of defining it. The important thing is, however, to adopt one definition and then apply it consistently throughout the analysis and for the comparisons with the benchmark(s).

(iii) The *stability of the clustering* is another potential outcome. If a map is trained on one data set you get one result. But would a similar map result if trained on different data samples? What happens if noise is added or reduced in the data? What is the impact of adding or deleting outliers? How stable is the number of clusters against variations of the cluster threshold? In using supervised neural nets we tested models on the basis of several out-of-sample data sets as well as on the basis of several random-selected samples of historical data. Testing of unsupervised neural networks should be done similarly: on multiple data samples, some with more noise, some with less noise, some with and some without outliers in the data. A comparison of the results would then provide a better selection of which map is more suitable for a particular use.

15.1.13 Summary

The quality of an unsupervised neural net model can and should be measured on the basis of (i) the number of clusters; (ii) the quality of clustering; (iii) the stability of clustering (as measured by the similarity or lack of similarity obtained by varying the testing data set). If we assess unsupervised neural net models in this way we are likely to find that there are many tradeoffs between quantity, quality, and stability of the clusters obtained by SOM. It would then be up to the user to determine what is the best combination in the light of the desired objectives of the study. Some applications may demand maximum data reduction (minimum number of clusters), and are happy with coarse map quality and low stability; other applications may demand refined maps (i.e. sharp differences between clusters) and good stability, but do not require a lot of data reduction. For example, in macro-economic analyses, analyses of world development indicators, environmental conditions, analyses of global poverty and the like, maximum data reduction may be most desired because the maps would be mainly used for policy formulation and macro decision-support. In other applications such as mapping opportunities for options and future trading, fund manager selection, client segmentation, product differentiation, or market analyses, much finer differentiation between clusters may be desired.

There is a vast domain of research and innovation to be done in this area, in particular in developing standards and a standard method for measuring the value added of the SOM in financial, economic and marketing applications.

15.2 Sample Application on Country Credit Risk Analysis

In Chapter 3 of this book we focused on the similarities and differences between mutual funds investing in overseas markets. In Chapter 6 we looked specifically at the investment opportunities of emerging markets. In this section we will focus on the risks involved in investing in various stock markets around the world. We use SOM to analyze these risks and to discover groupings of countries with similar risk patterns.

The analysis is based on "Global Investing: The Game of Risk" prepared by Greg Ip for the Wall Street Journal of June 26, 1997. In his article Greg Ip ranked 52 countries in the world based on their economic performance; political, economic and market risks; the depth and liquidity of their stock markets; and the regulation and efficiency of their markets. As a result of the analysis countries were classified into five groups: (i) those most similar to the US; (ii) other developed countries; (iii) mature and emerging markets; (iv) newly emerging markets; and (v) frontier markets. The countries classified in each group were:

- Group 1 – most similar to the US included the United States, Australia, Canada, Denmark, France, Germany, Ireland, Netherlands, New Zealand, Sweden, Switzerland and United Kingdom.
- Group 2 – other developed countries included Austria, Belgium, Finland, Hong Kong, Italy, Japan, Norway, Singapore and Spain.
- Group 3 – mature and emerging markets included Argentina, Brazil, Chile, Greece, Korea, Malaysia, Mexico, Philippines, Portugal, South Africa and Thailand.
- Group 4 – newly emerging markets included China, Columbia, Czech Republic, Hungary, India, Indonesia, Israel, Poland, Sri Lanka, Taiwan and Venezuela.
- Group 5 – frontier markets included Egypt, Jordan, Morocco, Nigeria, Pakistan, Peru, Russia, Turkey and Zimbabwe.

Note that in Ip's analysis one country, the US, was used as a basic benchmark for classification of countries; also, the grouping he derived divides the countries into five groups of about equal size. Note also that the criteria used by Ip for grouping is not explicitly provided in the article; furthermore, countries with so-called frontier markets had a lot of missing data (only three countries out of nine in this group have data for most indicators).

We will use the data provided by Ip in his Wall Street Journal article to conduct a SOM analysis. To do this we will follow the steps outlined above for creating visual displays and clustering of the country risks of these 52 countries. To facilitate this task we use Viscovery SOMine from Eudaptics GmbH which allows to undertake the analysis in an interactive and intuitive way.

Step 1

Define the purpose of the analysis

The purpose of our analysis is to verify the country risk analysis presented in the Wall Street Journal of June 26, 1997 and to produce a better rationale for groupings and visual presentation of the data. A secondary objective is to produce summary data on each group (which is not available in Ip's article) and to be able to look up missing values of indicators, in particular for the countries in so-called frontier markets.

Step 2

Select the data source and quality

Greg Ip's analysis was based on country data published in the Wall Street Journal which used as sources the Dow Jones, Morgan Stanley, Standard & Poor, Moody's Investment Services, IFC, World Bank and IMF. All sources of data are reputable and we can safely say that the data is of high quality (assuming no errors are made in copying or printing the data).

Step 3

Select the data scope and variables

The tables presented in the Wall Street Journal contained the following indicators for 52 countries:

- Compounded Total Annual Return over 3 years (in US$ as of 3/31/97)
- Price Earnings Ratio (versus fiscal 1996 earnings per share as of 4/17/97)
- Price Earnings Ratio Forward (as of 5/1/97)
- Historical Earnings Growth (5-year annual average % growth as of 4/17/97)
- Projected Earnings Growth (average annual % growth next 5 years as of 4/17/97)
- Dividend Yield (as of 4/17/97)
- GNP/capita 1995 (US$ 1995)
- Real GDP Growth (average 1990–95 in percent)
- Projected GDP Growth 1997
- Projected Inflation Rate 1997
- Short Interest rate (May–June 1997)
- Market Capitalization (in millions of $ as of 12/3/96)
- Turnover % (1996 trading volume as % of market capitalization as of 12/31/96)
- Total number of Companies listed (as of 12/31/96)

- Volatility (annualized standard deviation, 5-year average as of 3/97)
- Correlation of the market versus US (5-year average)
- Settlement Efficiency (trade size, failed trade frequency, out of possible 100, 1st quarter 1997)
- Safekeeping Efficiency (dividend collection, shareholder rights out of possible 100; 1st quarter 1997)
- Operational Costs (average annual impact in basis points, 1993–96)
- Age of Market (1997 minus year stock exchange established).

We did not use any of the indicators that were measured in local currencies, nor the credit ratings from Standard and Poor which were expressed on a scale from AAA to B. For the first SOM we limited the data scope to the economic performance measures: GNP per capita, GNP growth, projected GNP growth, inflation rate, and short-term interest. Subsequently, we expanded the data scope to the market-related indicators, e.g. market capitalization, number of companies listed, PE ratios, turnover, volatility, etc.

To start the analysis we first select the data source: Viscovery reads data sources stored in a text file or Excel spreadsheet. All of the data listed in Ip's article was entered in an Excel spreadsheet. This input data file was organized in rows and columns with each row containing all the data on a specific country, and each column containing the indicators listed above. This input data file also contained in the first row an abbreviated indicator description and in the last column of the spreadsheet the labels or abbreviated names for each country.

If no indicator names are specified, the components are named Component 1, Component 2, etc. Missing data are indicated by "." or "x".

To create our first map, we clicked on New Map in the File menu. The New Map wizard appears. This wizard guides the user through several steps until all information necessary to create the map is collected.

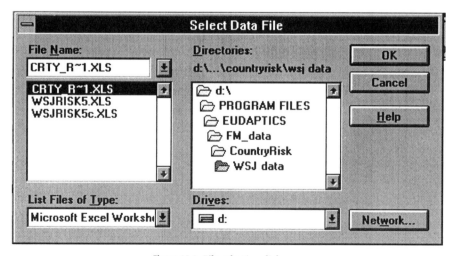

Figure 15.1. File selection dialogue.

The Select the Data File dialog appears where we choose the .xls or a .txt file in which the data is stored (Figure 15.1). By clicking OK and Next we proceed to the next window.

In the next window, called the Component window, you can select the data variables to be included in the analysis by placing a check mark in front of the components (variables, indicators and components will be used synonymously hereafter to refer to the columns in the data file). In this way one data file can be used for multiple analyses by using various subsets of the data, without the user needing to create dedicated data files for each analysis (as is the case in SOM_PAK). The Component window shows also the minimum and maximum of each variable (Figure 15.2). When all variables to be used have been selected you can choose to preprocess the data by clicking on Preprocessing or to click the Next button that leads to the next window.

Step 4

Decide how each of the variables will be preprocessed

If Preprocessing is clicked the Data Modification window (see Figure 15.3) appears. In this window various preprocessing options are provided. The main ones are (i) data scaling, (ii) variable priority, (iii) data modifications, and (iv) data transformations.

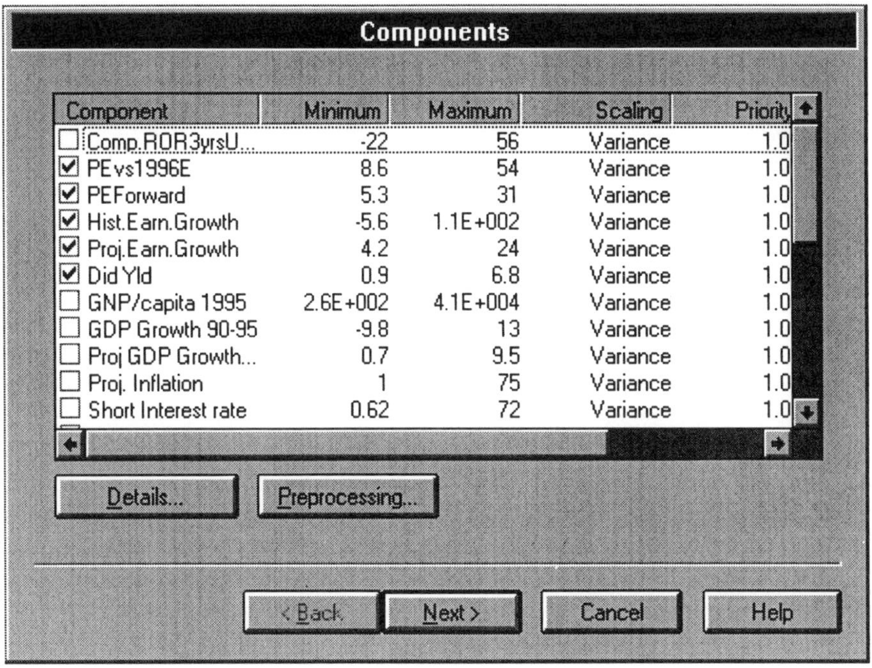

Components

Component	Minimum	Maximum	Scaling	Priority
☐ Comp.ROR3yrsU...	-22	56	Variance	1.0
☑ PEvs1996E	8.6	54	Variance	1.0
☑ PEForward	5.3	31	Variance	1.0
☑ Hist.Earn.Growth	-5.6	1.1E+002	Variance	1.0
☑ Proj.Earn.Growth	4.2	24	Variance	1.0
☑ Did Yld	0.9	6.8	Variance	1.0
☐ GNP/capita 1995	2.6E+002	4.1E+004	Variance	1.0
☐ GDP Growth 90-95	-9.8	13	Variance	1.0
☐ Proj GDP Growth...	0.7	9.5	Variance	1.0
☐ Proj. Inflation	1	75	Variance	1.0
☐ Short Interest rate	0.62	72	Variance	1.0

Details...   Preprocessing...

< Back   Next >   Cancel   Help

Figure 15.2. Component window and selection of data variables.

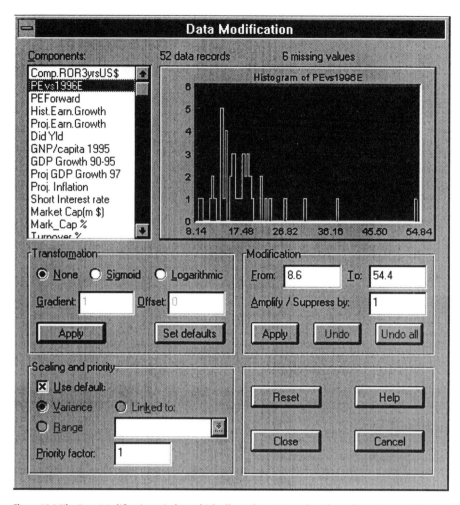

Figure 15.3. The Data Modification window which allows the user to select the scaling, data modification and transformations.

The map-creation process always works with an internal representation. This internal representation defines the topology of the data distribution, and thus the organization of the final map. In Viscovery, the user has several options to influence this internal representation, as outlined below. After having chosen some options, he or she needs not take care of the settings any more. Viscovery handles all scaling factors, priorities and transformations internally and interacts with the user only in terms of original data units. To keep track of the actions applied, they are reported in the Components window and in the Properties window of the map.

Data scaling
The user has four options to influence the scaling of components.

- Scale on the basis of variance: by selecting Variance scaling, the active component is internally scaled by variance, i.e. divided by its standard deviation.
- Scale on the basis of range: by selecting Range, the active variable is divided by its value range. If VS proposed Variance scaling and the user manually changes the scaling to Range, he or she is practically setting a higher priority factor. The relative impact of this variable on the map-creation process will then normally be increased.
- Scale on the basis of Linking to other variables: if two components are both measured in the same units, e.g. meters, and both are of the same range, it may be useful to scale both parameters with the same factor. VS provides a simple option to link the scaling of one variable to the scaling of another.
- Scale by default: in the default mode a component is automatically scaled by Variance if its range is smaller than 8 times the standard deviation; otherwise the scaling is set to the data Range. This heuristic number has in practice been shown to yield fairly natural scaling and has been introduced for convenience. In most cases it is best to let Viscovery determine the scaling by default.

Variable priority

The variable Priority gives an additional weight to a component by multiplying its internal scale by that factor. If the Priority is set greater than one, this variable will internally extend over a wider range so that clusters along that axis will become more lengthy. In contrast, if the Priority is set to a number less than one, then this component will be squeezed, thus becoming less relevant for the map-creation process. As a special case, if the Priority factor is set to a very small value, e.g. 0 to 0.1, then that component becomes irrelevant for the mapping process. This can be used to simply associate a parameter to the rest of the data without giving it an impact on the ordering process itself. For example, in the analysis of the country risk data, if we want volatility to be a key influence on the display, we could set the Priority of volatility to a value greater than one. Likewise, if we think that the age of a market does not have a major impact then we could set the Priority for Age to a smaller value. Obviously, setting Priority factors can cause significant changes to the resulting map.

Data modification

By selecting a range for a variable in the histogram and setting an amplification factor, it is possible to add or remove data records in the specified range of the data space. Amplifying ranges is a delicate action and must be handled with care. Less critical, data suppression can be helpful in eliminating outliers or trimming the scope of a variable's histogram. If the user is interested in particular ranges of the data, an amplification can be an efficient means to focus the map creation on this range. Any modification of the data set will not only be reflected in the histogram of the specified component but also change the histograms of other components, because whole data records, not only components, are added or removed.

Data transformations

By applying a transformation you can influence the density characteristics of a variable's distribution. The two main transformation types are logarithmic and sigmoid. For convenience, default parameters for these two transformations are provided (suggesting plausible choices for gradient and offset).

Applying a data transformation redefines the internal representation of a variable with the specified function. As this changes the distances between records, this has an impact on the intrinsic neighborhood relationships within the data set. Therefore, this feature should be used with caution.

Suppose that the records of a specific variable are mostly concentrated at the lower end of the range of its histogram and fewer records have higher values. In this case the user may wish to start the map-creation process with a more uniform density distribution. Applying a logarithmic transformation would equalize the distribution, since the logarithmic function gives a higher "resolution" to small values in the histogram. The smaller values would then gain stronger impact on the clustering of the data. In contrast, a sigmoid function can create a more balanced distribution by stretching the center of the histogram and squeezing the ends. Sigmoid transformations are a good way to handle outliers without discarding them.

Since in the country risk analysis the market capitalization of a few countries is vastly larger than that of others, a log transformation is applied on the market capitalization as well as on the number of companies listed on each market, which may improve the resulting display and clustering of countries.

In our sample application on country risk we chose to scale all the data by default (which was the variance in all cases) and we applied no data modifications or transformations. Furthermore, all variables were kept at the same priority level. The Data Modification window, shown in Figure 15.3, was thus left in the default mode.

Step 5

Select the clustering and visualization method(s); consider the use of hybrid methods

Since the objectives of this analysis are to verify a particular grouping of countries based on risk indicators and to demonstrate how to apply a user-friendly tool for creating SOM displays, we did not use any other statistical techniques or any hybrid combination of SOM with other approaches. For future analyses it would, however, be desirable to be able to combine SOM with other techniques, e.g. genetic algorithms, fuzzy logic or PCA analysis.

Step 6

Determine the desired display (map) size and shape, and the degree of detail that is desired

After closing the preprocessing window we click Next in the Components window to reach the Target Map window (see Figure 15.4). This window enables the user to specify basic map parameters. These include:

- the number of nodes or size of the map – the nodes are organized in a hexagonal grid. After specifying an approximate number of nodes in the map, the final

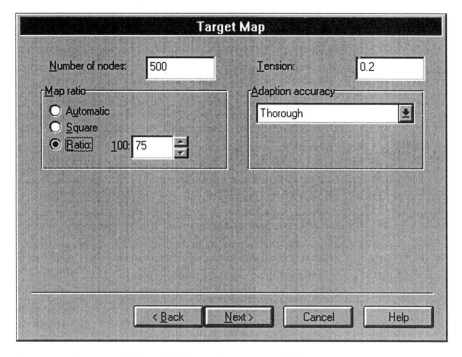

Figure 15.4. The Target window which allows the user to specify the map, the map ratio, the tension and the adaptation accuracy.

number of rows and columns in the resulting map will be automatically determined. We choose 500 nodes (or about ten times the number of input vectors).

- the map ratio – this value describes the relation between the width and the height of the map. The automatic setting causes Viscovery to compute the map ratio from the input data. The format of the map will then provide the best fit to the source data set. Alternatively the ratio can be specified by the user. We choose a ratio of 100 to 75.

- the map tension – the "tension" is the reach of the neighborhood function in the end of the training process. This value measures the ability of the map to adapt to input data. The default of 0.5 is normally a good choice. We lowered the default to 0.2 in order to obtain a map that would show greater detail. Values higher than the default would give a coarser map.

- the map-creation accuracy – this value can be set to Fast, Normal, Thorough, or Slow. These are pre-defined learning schedules specifying a set of internal parameters which affect the map creation.

After the selection of the map parameters in the Target Map window the Next button brings up the Properties window which provides a summary of the important map parameters to be used (see Figure 15.5):

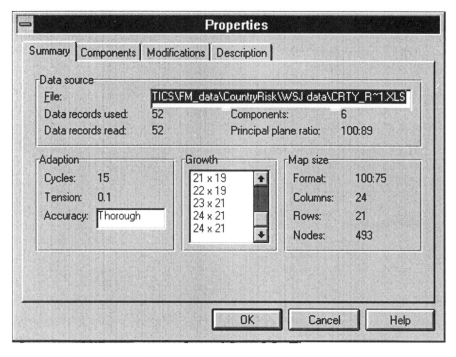

Figure 15.5. Summary of the map parameters selected for the creation of a new map.

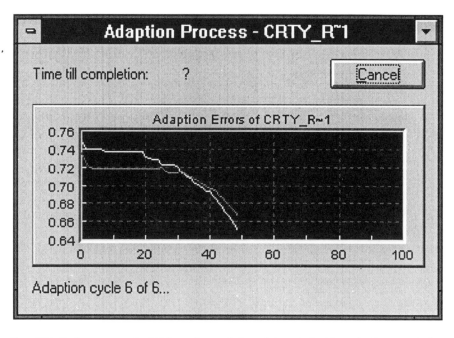

Figure 15.6. During map creation VC keeps the user informed through the Adaptation Progress window.

- the data source – the name of the source data file and the number of records read and used, respectively;
- the map size – the format of the map, the number of columns and rows and total number of nodes;
- the adaption schedule – the intermediate map sizes during the adaption cycles, resulting from the chosen format, tension and accuracy.

When you have verified all these map parameters clicking OK starts the map-creation process. The Adaptation Progress window (see Figure 15.6) appears on the screen showing the progress in creating the new map and two important error measures associated with the training process. When the process finishes the Cluster window appears. This gives a first view of the map created from the input data (see Figure 15.7).

Step 7

Use relevant sample data in training the map

In this example on country risk analysis we used all 52 records for creating the display. We recall that six records had a lot of missing data. One of the nice features of the SOM algorithm is that it can handle input vectors with missing data. Thus, the fact that several values were missing from six records in the so-called frontier markets does not require that we delete these records. When a lot of values are missing in particular inputs, say more than 50% missing, then you may prefer to keep these input vectors apart for testing after the map has been trained.

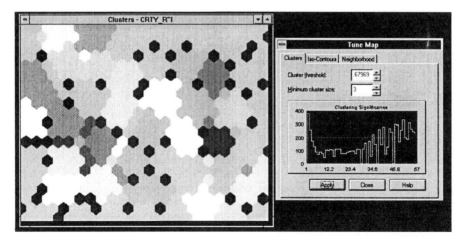

Figure 15.7. Sample output from the adaptation process. The window on the left shows the SOM; the window on the right shows the Tuning window.

Step 8

Tune the map for optimal clustering and visualization

Once you have created a new map it can be investigated in several ways:

- viewing different aspects of the map in separate windows;
- showing or hiding various parts of a map in each window;
- investigating values of individual nodes or clusters;
- identify or paste labels for individual nodes or clusters.

Viewing different aspects of the map

In Viscovery the following menu commands are available to open separate windows, referred to as Map windows. Each of them displays different aspects of the map:

- The Clusters command from the Analyze menu opens a Cluster window. This window shows by default when a new map is created. The Cluster window shows the map partitioned into disjoint regions, or clusters, which are usually separated by a dark-gray area, the separating area. The partitioning can be influenced (called clustering, in the Tune Clusters window of the Tune Map window). You can specify different values for the cluster threshold and the minimum cluster size. The Cluster window can alternatively show the U-matrix view of the map and similarity shadings (so-called Iso-contours).

- The Components command from the Analyze menu opens one or more Component windows. This command brings up the Show Components dialog, where the user can select the components to display. Each Component window visualizes the values of one particular component – this is the projection of the component onto the map. The color scale in the lower part of the window helps match colors against values.

- The Frequency command from the Analyze menu opens a Frequency window. This window indicates how many data records from the source data set matched each node. Nodes that did not match any data record are white. Nodes that did match at least one data record show a shade of red that indicates the number of matches. The color scale in the lower part of the window helps match colors against numbers.

- The Quantization Error command from the Analyze menu opens a Quantization Error window. The quantization error is a measure of how well the data vectors from the source data set matched a specific node. The average of the values over all nodes is the quantization error of the map. The quantization error is displayed as white at nodes where no data records matched (i.e. where the frequency is zero). Non-zero quantization errors are displayed in shades of green. The color scale in the lower part of the window helps match colors against error values.

- The Curvature command from the Analyze menu opens a Curvature window. As the map is a two-dimensional surface that approximates the data distribution of the source data set (which is usually of higher dimensionality), it (usually) cannot be a plane but must bend through the data space, thereby forming curves, saddles, etc. The Curvature window displays how strongly the map (surface) is bent in each node. The color scale in the lower part of the window helps match colors against curvature values.

Showing or hiding various parts of a map

Independently for each window, the user can show and hide various highlights using menu commands and accelerators.

- Show or hide labels using the Labels command.
- Show or hide separators using the Separators command.
- Show or hide iso-contours using the Iso-Contours command.
- Show or hide the neighborhood using the Neighborhood command.

Step 9

Interpret the results, check the values of individual nodes and clusters

Viscovery provides several means to study quantitative features of the map. The Values window offers a way to inspect component values at single nodes as well as values computed from a collection of nodes over certain areas of the map. By choosing Values from the Analyze menu the Values window appears.

If Node is entered in the Range box then clicking on any node of the map will show the values associated with that node in the Values window. If Cluster is entered in the Range box, component values for an entire cluster are provided. Clicking into any cluster will show the values for each component computed over that cluster. This feature makes it possible to identify clusters with distinguished component values interactively.

The Values window displays the mean, standard deviation, minimum, maximum and sum for a single node or a collection of nodes. The node for which information is displayed can be selected manually by clicking on the node in any map window, or if there is ongoing monitoring (and it is not paused), the current monitor point defines the node.

Figure 15.8 provides examples of the Values window. At the top of the figure we see the information related to a node in the credit risk map that includes the US; at the bottom of the figure we see the information of a node in the credit risk map that groups some of the ASEAN countries. The difference between both clusters is obvious. The latter have lower PE ratios, lower project earnings growth, much lower market capitalization and number of companies listed, but much higher volatility than the US markets.

This feature of Viscovery to inspect the values of individual nodes and clusters provides an efficient way to investigate the results of a particular display and allows to tune a map by changing the cluster threshold or size.

Values - CRTY_R~1				
Range: Cluster X: 0 Y: 2 Nodes: 8				
Component	Mean	Standard de...	Minimum	Maximum
PEvs1996E	19	0.077	19	20
PEForward	16	0.051	16	16
Hist.Earn.Growth	26	0.048	26	26
Proj.Earn.Growth	14	0.059	14	15
Did Yld	2	0.0019	2	2
Market Cap(m $)	8.5E+006	7.9E+004	8.2E+006	8.5E+006
Turnover %	93	0.052	93	93
No Companies	8.5E+003	6	8.5E+003	8.5E+003
Volatility	7.4	0.23	7.3	8
Frequency	0.13	0.33	0	1
Quantization error	2.4E-020	6.5E-020	0	2E-019
Curvature	0.66	0.37	0.071	1.1

a

Values - CRTY_R~1				
Range: Cluster X: 6 Y: 5 Nodes: 309				
Component	Mean	Standard de...	Minimum	Maximum
PEvs1996E	17	3.4	9.1	27
PEForward	14	2.5	6.7	19
Hist.Earn.Growth	18	11	0.6	41
Proj.Earn.Growth	14	3.2	8.4	24
Did Yld	2.3	0.64	1	4.1
Market Cap(m $)	2.1E+005	2.8E+005	1.8E+003	1.7E+006
Turnover %	51	29	7	1.3E+002
No Companies	4.7E+002	4.7E+002	64	2.4E+003
Volatility	21	7.9	10	36
Frequency	0.097	0.3	0	1
Quantization error	7.9E-016	1.2E-014	0	2E-013
Curvature	0.58	0.2	0.0089	1.2

b

Figure 15.8. a and **b**: The values of two clusters of a newly created map on country risk.

Step 10

Define or paste appropriate map labels

Each node in the map can carry one- or multiple-line labels. Labels can be pasted onto the map by selecting them from the original data file (that is from the last column of the data file) and pasting them onto the map, or the labels can be entered individually by selecting a place in the map and typing a label name. Labels are saved with the map and can be edited.

Figure 15.9 shows the credit risk map we created with labels. Note that each label in this case is composed of two lines including the original Ip group number and the abbreviated country name.

While the appropriate labeling of a map will clarify the groupings or clusters obtained through SOM, there is no better way to find the non-linear relationships or dependencies between the various indicators than by showing the Component windows. By placing a series of Component windows side by side next to the main map (see Figure 15.10) you can directly investigate the dependencies between components. For example, in Figure 15.10 the component windows of the current Price Earning ratio, forward Price Earning ratio, dividend yield, market capitalization,

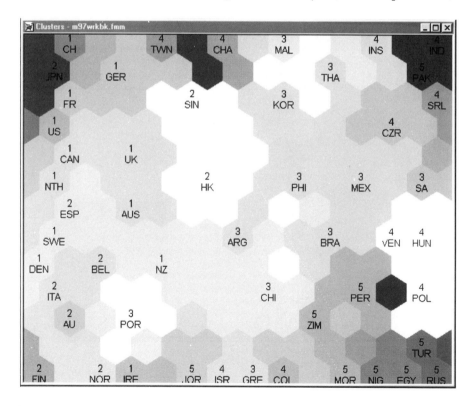

Figure 15.9. SOM map with labels on country credit risks derived from data published in the Wall Street Journal of June 26, 1997. *(This figure can be seen in color in the Color Plate Section.)*

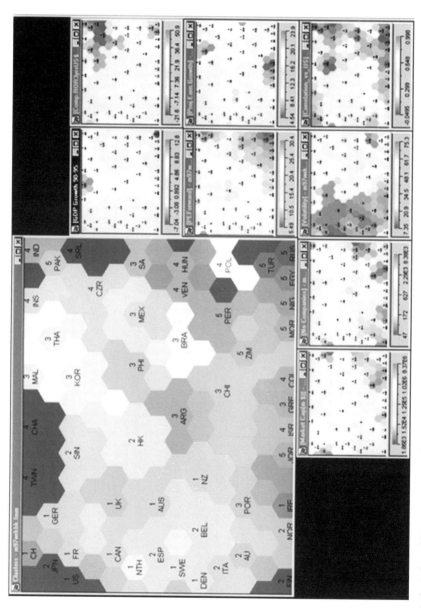

Figure 15.10. Typical window of Viscovery SOMine showing a SOM map with eight component planes allowing to derive non-linear dependencies among variables contributing to the clustering of countries based on country risk measures. Data used for this map was published in the Wall Street Journal of June 26, 1997. (*This figure can be seen in color in the Color Plate Section.*)

number of companies and volatility are shown. For each of these component windows the color bar at the bottom shows the range of values for each component. Lower values for each component are in blue and higher values are in red; values in between go from light blue to green to yellow to orange.

By comparing component values in certain areas the user can read off non-linear dependencies and thus visually identify the meaning of clusters. For example, it can be seen that markets in the top right corner of the map have the highest volatility (Turkey and Poland); those in the top left corner have the lowest volatility (US); the projected earnings growth is lowest for Israel, Columbia and Greece, and highest for Hungary and Thailand; the number of companies listed is highest in the US and India.

One of the objectives of our analysis was to be able to look up values. An advanced feature of Viscovery is the association of values with data records. One way of doing that has already been described: just set the priority of a specified component to zero during training. The association of new components can, however, also be done explicitly after the map has been trained.

For example, in the analysis of country risk we could turn off the Priority for Volatility by setting it to 0. This means that the Volatility is not taken into account during the map-creation process and is, therefore, an associated value. The Component window of the Volatility can then be compared with the others to find those values that result in a high Volatility, for example. Figure 15.10 shows that the associated Volatility is nicely separated in an area with high values and an area with low values. This means that the Volatility can be reasonably deduced from the other values. If high and low values were randomly distributed over the whole map, we would have to conclude that there is no relationship between the Volatility and the other country risk variables.

If you associate a value with data records representing a time series, you get a prediction of that value. In sum, Viscovery can also be used for predicting market volatility, PE ratios, inflation and/or stock market performance indices.

Steps 11 and 12

Produce a summary of results that highlight the differences between clusters and evaluate results

The country risk map created with SOM produces quite different results than those reported in the Wall Street Journal.

If we artificially impose five clusters as the Wall Street Journal article did, we obtain a map with a different grouping of countries (map not provided here). A credit risk map that is artificially limited to five clusters shows the following grouping (with group numbers and order having no particular significance):

- Group 1 – US
- Group 2 – India
- Group 3 – Japan
- Group 4 – Turkey and Poland
- Group 5 – all other countries.

A map that artificially constrains the number of clusters shows the US, India and Japan in separate clusters. Both the US and Japan have very large market capitalization; India has a very large number of companies listed on its exchanges (one of the largest numbers of companies after the US exchanges). Turkey and Poland stand out as a group and all the other countries are undifferentiated. Clearly, the artificial limitation of SOM to five clusters does in this case not provide a great deal of new information.

If the artificial constraints are removed we obtain a completely different grouping of countries, based on country risks. From Figures 15.9 and 15.10 we find the following clusters (neither cluster numbers nor order have any meaning):

- Cluster 1 – Australia, New Zealand, Canada and most of Europe
- Cluster 8 – most of Latin America, and Eastern Europe
- Cluster 6 – Mexico, Philippines, South Africa and Czech Republic
- Cluster 3 – Korea, Malaysia, Thailand and Indonesia
- Cluster 2 – Singapore and Hong Kong
- Cluster 4 – Hungary and Venezuela
- Cluster 7 – Brazil
- Cluster 5 – Poland
- Cluster 9 – India and Pakistan.

Cluster 1 groups most of the developed stock markets. Cluster 2 identifies two Asian markets which are larger and more active than all the others; cluster 3 groups most of the ASEAN markets; clusters 4 and 5 identify markets that either have been very volatile or have moved ahead of others in the emerging market category. Falling outside these clusters would be the US, Japan, China, Taiwan, Sri Lanka and Finland.

To facilitate comparisons we retained the clustering number closest to the first SOM. In consequence, clusters 1 to 5 are very close to the first five clusters of the previous SOM; stock markets of countries listed in clusters 6 to 9 form separated clusters. Thus we find that cluster 6 includes Mexico, Philippines, South Africa and Czech Republic; cluster 7 is Brazil; cluster 8 includes most of Latin America (except Mexico and Brazil and countries in Eastern Europe); cluster 9 groups India and Pakistan.

Most important of all we note that if we take into account all of the data, perform appropriate transformation, and let the data self-organize that countries group quite differently than when we artificially impose five groups of approximate equal size, and use the US as a benchmark. *The SOM map we obtained does not correlate with the Wall Street Journal analysis of country credit risks grouping. There is also a strong case from a domain expert perspective for the argument that the SOM map results provide a more logical grouping than the original division of the 52 countries into five groups with approximate equal number of countries.*

This example proves several things.

- The value of data transformations and modifications: without log transformations of the market capitalization and number of companies listed the US and Japan would capture two clusters, and Singapore and Hong Kong would not be singled out from among other markets in Asia.

- If no constraints are imposed on the number of clusters, the cluster significance (measure used in VS to optimize the number of clusters) drops significantly after nine clusters. Thus, nine clusters is a better representation of the data than fewer than nine clusters.

- A SOM map which provides for self-organization of the data results in quite different groupings of countries based on credit risks than a grouping based on comparisons with a single country and (undefined) criteria.

This example also demonstrates that SOM should be used only in applications where the user also understands the data. Knowledge domain expertise is essential and should always play an important role in making an effective SOM application.

15.3 Conclusions

This chapter has outlined a procedure for data mining using self-organizing maps. We have illustrated this procedure on the basis of an analysis of country risks. We showed with a practical example how to apply the procedure outlined in Section 15.1 and what the advantages are of using commercially available, user-friendly software to facilitate the creation of self-organizing maps. As indicated throughout this book, *a lot remains to be done in terms of using the SOM algorithm in finance, economics and marketing. We hope that the introduction provided in this book will entice you to look into visual data explorations with self-organizing maps.*

Notes

Introduction

[0.01]   In statistics it is customary to call the components of the data vectors *observations*. They may also be called *components*, according to the mathematical terminology, or *features*, which is a commonly used in the pattern recognition literature. In econometric studies the variables are often called *indicators*. In various applications, the names *descriptors* and *attributes* also appear.

[0.02]   Ripley (1996) divides unsupervised statistical data-analysis methods into clustering methods, projection methods, and multi-dimensional scaling (MDS) methods. Differentiation between the latter two is not useful here, since for patterns represented as metric vectors the MDS methods essentially form non-linear projections.

[0.03]   Another method for displaying high-dimensionality is *projection pursuit*. In exploratory projection pursuit the data is projected linearly, but this time a projection which reveals as much as possible of the non-normally distributed structure of the data set is sought. This is made by assigning a numerical "interestingness" index to each possible projection, and by maximizing the index. The definition of interestingness is based on how much the projected data deviates from normally distributed data in the main body of its distribution. After an interesting projection has been found, the structure that makes the projection interesting may be removed from the data, after which the procedure can be restarted from the beginning to reveal more of the structure of the data set.

[0.04]   There are a multitude of variants of MDS with slightly different goals and optimization algorithms. The first MDS for metric data, i.e. data for which the exact values of the distances between data items are meaningful and important, was developed in the 1930s. The methods were later generalized for analyzing non-metric data, data for which only the order of the distances between the data items is important. Even the common structure in several dissimilar matrices corresponding to, for instance, evaluations made by different individuals can be analyzed.

[0.05]   There is also a computationally efficient method that may be viewed as a modification of the Sammon's mapping, called *curvilinear component analysis* (Demartines, 1994; Demartines and Hérault, 1997). The data items are first clustered and then the clusters are mapped with a fast algorithm in which the relative emphasis on the large and small distances can be regulated.

Chapter 2

[2.01]   Cox et al. (1985a) or arbitrage-free condition in Vasiceck (1977).
[2.02]   Longstaff and Schwartz (1992) assume that the price of risk does not depend on time.
[2.03]   For more information on those theoretical aspects, see Cottrell et al. (1997).

[2.04]   For example, the antithetic or contravariate method (Boyle, 1977) or, more recently, the Martingale Variate Control.

[2.05]   The non-parametric approach that is used in this chapter was first proposed by Cottrell et al. (1996a) and Cottrell and de Bodt (1997); see also Cottrell et al. (1997).

[2.06]   The number of units chosen for the classification of interest rates structures is problem dependent. A good criteria to validate the choice is the observation of the evolution of the ratio Intra-classes Sum of Squares to Total Sum of Squares as the number of units increases.

[2.07]   We will not reproduce all the statistical results here but focus on the more important results.

[2.08]   Table 2.1 presents the variance explained by the first four factors and the correlations between the first two factors and the level of the short rate and the spread (difference between the long rate and the short rate).

Chapter 3

[3.01]   The maps shown in this chapter were produced using the SOM_Package, a public domain software developed by the Neural Network Center at the Helsinki University of Technology. The selection of "best" maps was done on the basis of multiple simulations using different map initializations, different neighborhood functions, map sizes, and/or training procedures. A detailed explanation of the SOM methodology and the options imbedded in the SOM_Package can be found in Part 2 of this book (Kohonen et al., 1995).

Chapter 4

[4.01]   The SOM can also be utilized in the subsequent phases of quantitative analysis, but we shall not pursue that subject further here.

[4.02]   Reporting the training set accuracies may here be justified by noting that in our other studies we have found overlearning not to be a problem of any practical significance for this small a map, when there is as much data available as in the present study.

Chapter 5

[5.01]   Expert, No 24, July 30 1997, p.6.

[5.02]   Russian Banks Accounting Information for 1994, Central Bank of Russia, 1995.
Russian Banks Accounting Information for 1995, Central Bank of Russia, 1996.

[5.03]   Further increase of parameter dimension makes visualization impossible

[5.04]   We used Oja's (1994) network with 30 input and 10 output neurons as a raw data preprocessor to achieve a three-fold compression without any significant loss of information.
A neuron is represented by its weight embedded in the input space, so the distances between inputs and neurons are well defined..

Chapter 6

[6.01]   The World Bank ranks economies based on per-capita gross domestic product. The current World Bank classification of countries includes:

- low-income countries, which are those with less than $695 per-capita GDP;
- medium-income countries, which are countries with $695 to $2785 per-capita GDP;
- medium-high income countries, which are countries with $2785 to $8626 per-capita GDP;
- high-income countries, which are countries with over $8626 per-capita GDP.

Based on the criterion of a per-capita GDP of less than $8626 per year, there are approximately 170 countries around the world that meet the current World Bank definition of "emerging markets". There is, however, a significant problem with defining emerging markets on the basis of per-capita GDP: per-capita GDP values are averages that do not show the degree of income inequality. For example, for a long time the World Bank has published a GDP per capita for China that was way below what many economists and professional China observers estimated the GDP per capita to be. Another important shortcoming is that GDP per capita does not tell us anything about which countries have equity markets; whether those that have an exchange have a regulated and functioning exchange. Even if there is a functioning exchange, regulated exchange shares traded on the exchange may be restricted to local investors and not available for purchase by foreign investors. Thus, of 170 countries that are defined by the World Bank as "emerging markets" there are only about 80 countries that have functioning securities exchanges.

[6.02]   In the period 1989–1995 the emerging markets grew at an average annual rate of 4.9%; the developed countries grew at an average annual rate of 2.9%. The cumulative growth of the emerging markets over the past seven years was 33.5%, whereas that of the developed countries was 18.6%. The earnings of all companies included in the IFC (Investable) index of emerging markets grew at an average of 9.7% over the past six years compared with 4% in developed countries. The book value of emerging market companies grew at 13.8% compared with 9.2% in developed countries. The dividends of emerging market companies increased at an average rate of 7% compared with 6.7% in developing countries.

[6.03]   In 1995 access to the IFC Emerging Market Database was facilitated through a Windows-based data retrieval and graphing system. The Emerging Markets Database is available from IFC in computerized form and as well as on hard copy. Publications from IFC on this subject include the *Monthly Review of Emerging Stock Markets*, the annual *Emerging Stock Markets Factbook*, the monthly *Constituents of the IFC Indexes* and *The IFC Indexes: Methodology, Definitions and Practices*.

[6.04]   Survey of Banking in Emerging Markets, in The Economist, April 12, 1997

Chapter 12

[12.01]   WEBSOM home page (1996) available at http://websom.hut.fi/websom/.

Glossary

Adaptability The ability of computing systems (neural or otherwise) to adapt themselves to the data. Such an ability is synonymous with the ability to learn.

Adaptive fuzzy system A fuzzy system that forms its rules from data, or where a human expert does not tell the system what the rules are. Fuzzy rules are extracted from input–output data relations. An adaptive fuzzy system acts as a human expert in real time.

Adaptive resonance theory A two-level unsupervised neural network architecture which performs vector quantization with a variable number of models.

Antipersistence A reversal of a time series, occurring more often than reversal would occur in a random series. If the system has been up in the previous period, it is likely to be down in the next period, and vice versa.

Artificial intelligence (AI) AI views the mind as a type of logical symbol processor that works with strings of text or symbols much as a computer works with strings of 0s and 1s. In practice, AI means expert systems or decision support systems.

Attractor In a nonlinear dynamic process, a point of stability of its states. See also *Limit cycle, Point attractor, Strange attractor.*

Autoassociative memory A memory which is designed to transform an input pattern to itself. If the input pattern is noisy, degraded or incomplete the memory will still recall the original or learnt pattern.

Backpropagation A learning scheme by which a multi-layer feedforward network is organized for pattern recognition or classification utilizing an external teacher, and error feedback (or propagation).

BDS statistics Brock–Dechert–Scheinkman test for serial independence in a time series.

Beta The standardized covariance of an instrument with its complete class of similar instruments, i.e. the overall movement relative to its market.

Capital asset pricing model (CAPM) An equilibrium-based asset pricing model developed independently by Sharpe, Linter and Mossin. The simplest version states that assets are priced according to their relationship to the market portfolio of all risky assets, as determined by the securities' beta.

Chaos A deterministic, nonlinear dynamic system that can produce random-looking sequences of states of an infinite length. An aperiodic equilibrium state of a dynamical system. A system in a chaotic equilibrium seems to wander "at random" through states, yet the behavior is deterministic: a mathematical equation describes it exactly. If you know the equation and the present state exactly, you can predict any point of the chaotic path or trajectory. Chaos has the property that if you pick any two starting points of a chaotic system, no matter how close, they give rise to two paths that will diverge in time. Most dynamical systems have chaotic equilibria. These range from the interaction of subatomic particles to the bubbles in a hot tub and from the swirl of clouds in the sky to the distribution of galaxies in space. A chaotic system must have a fractal dimension and must exhibit sensitive dependence on initial conditions. See also *Fractal, Lyapunov exponent, Strange attractor.*

Coherent market hypothesis A theory stating that the market distribution may be determined by a combination of group sentiment and fundamental bias. Depending on combinations of these two factors, the market can be in one of four states: random walk, unstable transition, chaos, or coherence.

Competitive learning A learning rule where processing elements compete to respond to a given input stimulus. The winner then adapts to make itself more like the input. Different elements are adapting to different inputs.

Control strategy Specifies the order in which network weights are updated for digital implementations of learning functions.

Correlation dimension An estimate of the fractal dimension that (1) measures the probability that two points chosen at random will be within a certain distance of each other and (2) examines how this probability changes as the distance is increased. A dependent system will be held together by its correlations and will retain its dimension in whatever embedding dimension it is placed, as long as the embedding dimension is greater that its fractal dimension. White noise will fill its space because its components are uncorrelated, and its correlation dimension is equal to whatever dimension it is placed in.

Correlation integral The probability that two points are within a certain distance from one another is given by the sum or integral over fixed-size boxes in state space; used in the calculation of the correlation dimension.

Crossover A procedure or operator in genetic algorithms used to introduce diversity during reproduction. One-point crossover takes two child chromosomes, selects at random one point on those children, and swaps the genetic material on the children at that point. Two-point crossover works the same way but selects two points at random. The use of crossover is a very important feature of a genetic algorithm, and may be critical to its success.

Delta rule learning A type of learning where weights are modified to reduce the difference between the desired output and the actual output of a processing element. Synonymous for Robbins–Monro (Widrow–Hoff) learning. Realizes the minimization of mean squared error.

Divergence A situation where two indicators are not confirming each other. For example, in oscillator analysis, prices trend higher while an oscillator starts to drop. Divergence usually warns of a trend reversal.

Dynamic system A system the state of which changes with time. A simple version of a dynamic system is a set of linear simultaneous equations. Nonlinear simultaneous equations are nonlinear dynamic systems. In math a system described by a differential or difference equation – a system whose change is some function of time or of system parameters. In a broad sense everything is a dynamical system, the universe and all its components. The starting point of a dynamical system is an initial condition. The final point or points is the equilibrium state. In between lie the transient states. A dynamical system can have two types of equilibrium states: periodic and aperiodic. Aperiodic equilibria are *chaotic* or strange attractors. Once the system falls in one of these regions it moves around for ever, or until something bumps it into a new state, with no apparent structure or periodicity to the movement. The simplest periodic equilibrium is the fixed point attractor. There are also *limit cycle* attractors, where the state swirls round and round the same path. See also *Point attractor, Limit cycle, Strange attractor.*

Efficient frontier   In mean/variance analysis, the curve formed by the set of efficient portfolios – that is, those portfolios of risky assets that have the highest level of expected return for their level of risk.

Efficient market hypothesis (EMH)   A theory that states, in its semistrong form, that, because current prices reflect all public information, it is impossible for one market participant to have an advantage over another and reap excess profits.

Euclidean geometry   Plane or "high school" geometry, the first and perhaps most evident axiomatic treatment of geometry.

Expert system   A computer system that tries to simulate a human expert. A search tree and method of traversal in artificial intelligence. The expert provides her knowledge as if–then rules and a programmer codes these in software. Expert systems define a large logic tree or several small trees. The expert system has two parts: the knowledge base and the inference engine. The knowledge base is just the tree or trees of bivalent rules. The inference engine is some scheme for reasoning or "chaining" the rules. Fuzzy systems are a type of expert system since they too store knowledge as rules, but as fuzzy rules or fuzzy patches. Expert systems work with black–white logic and symbols. Fuzzy systems work with fuzzy sets and have a numerical or mathematical basis that permits both mathematical analysis and simple chip design.

Fault tolerance   A property of neural computing systems that allows the system to function and gradually degrade when a small number of processing elements are destroyed or disabled. See also *Graceful degradation*.

Feedforward network   A network in which information flow is all in one direction. In such networks there are no feedback loops from a processing element to a previous one.

Fractal   An object in which the parts are in some way related to the whole; that is, the individual components are "self-similar at all magnifications." An example is the branching network in a tree. Each branch and each successive smaller branching is different but all are qualitatively similar to the structure of the whole tree.

Fractal dimension   A classical way of describing a fractal dimension relates to measuring the length of a "wiggly" line with various different measuring sticks. For example, the coastline of Britain (like that of many other countries) is a wiggly line. The length of a coastline can be measured by counting how many times one can fit a measuring stick with a known length along the coastline and multiplying this number with the length of the measuring stick. While this seems to be a very obvious way it appears that the length of a wiggly line depends on the length of the measuring stick.

Fractal distribution   A probability density function that is statistically self-similar, i.e. in different increments of time, the statistical characteristics remain the same.

Fractal market analysis   A method to measure the memory in a time series, the fractal dimension and/or the correlation dimension in time series. See also *Rescaled range analysis, Fractal dimension, Correlation dimension*.

Fractional Brownian motion   A biased random walk; comparable to shooting craps with loaded dice. Unlike standard Brownian motion, the odds are biased in one direction or the other.

Fuzzy logic   Has two meanings. The first meaning is multivalued or "vague" logic. Everything is a matter of degree including truth and set membership. The second meaning is reasoning with fuzzy sets or with sets of fuzzy rules. This dates back to the first work on fuzzy sets in the 1960s and 1970s by Lotfi Zadeh at the University of California at Berkeley. Zadeh chose the adjective "fuzzy" over the traditional adjective "vague" in his 1965 paper "Fuzzy Sets" and the name has stuck.

Fuzzy rule A conditional of the form IF X IS A, THEN Y IS B where A and B are fuzzy sets. In mathematical terms a rule is a relation between fuzzy sets. Each rule defines a fuzzy patch (the product A × B) in the system "state space". The wider the fuzzy sets A and B, the wider and more uncertain the fuzzy patch. Fuzzy rules are the knowledge-building blocks in a fuzzy system. In mathematical terms each fuzzy rule acts as an associative memory that associates the fuzzy response B with the fuzzy stimulus A.

Fuzzy set A set whose members belong to it to some degree. In contrast, a standard or nonfuzzy set contains its members all or none. The set of even numbers has no fuzzy members.

Fuzzy system A set of fuzzy rules that converts inputs to outputs. In the simplest case an expert states the rules in words or symbols. In a more complex case a neural system learns the rules from data or by watching the behavior of human experts. Each input to the fuzzy system fires all the rules to some degree as in a massive associative memory. The closer the input matches the if-part of a fuzzy rule, the more the then-part fires. The fuzzy system adds up all these output or then-part sets and takes their average or centroid value. The centroid is the output of the fuzzy system. Each map from input to output defines one FLIPS or fuzzy logical inferences per second. The Fuzzy Approximation Theorem (FAT) shows that a fuzzy system can model any continuous system. Each rule of the fuzzy system acts as a fuzzy patch that the system places so as to resemble the response of the continuous system to all possible inputs.

Gaussian A system in which the probabilities of the states or outputs are well described by a normal distribution, or a bell-shaped curve.

Generalization The ability of a neural computing system to generalize from the input/output examples it was trained on to produce a sensible output to a previously unseen input. Compromise of the variance-bias dilemma.

Genetic algorithm (GA) Genetic algorithms are problem-solving techniques that solve problems by evolving solutions as nature does, rather than by looking for solutions in a more principled way. Genetic algorithms, sometimes hybridized with other optimization algorithms, are the best optimization algorithms available across a wide range of problem types.

Graceful degradation The ability of a neural computing system to only slowly cease to function properly if a small number of processing elements are destroyed or disabled. See also *Fault tolerance*.

Hausdorff dimension A different type of topological dimension which has the interesting characteristic that it can also take non-integer values. The Hausdorff dimension can be approximated by the fractal dimension or correlation dimension.

Hidden neuron Usually a nonlinear (or linear) processing element with no direct connections to either inputs or outputs. It often provides the learning capacity of the neural network.

Hurst exponent (H) A measure of the bias in fractional Brownian motion. $H = 0.50$ for Brownian motion; $0.5 < H < 1.0$ for persistent or trend-reinforcing series; $0 < H < 0.5$ for an antipersistent or mean-reverting system. The inverse of the Hurst exponent is equal to alpha, the characteristic exponent for fractal or Pareto distributions.

Layer The main architectural component of a neural network consisting of a number of processing elements of equal functionality and occupying a position in the network corresponding to a particular stage of processing.

Learning In the simplest form: self-adaptation at the processing element level. Weighted connections between processing elements or weights are adjusted to achieve specific results, eliminating the need for writing a specific algorithm for each problem. More generally: change of rules or behavior for a certain objective.

Learning schedule A schedule which specifies how parameters associated with learning change over the course of training a network.

Leptokurtosis The condition that a probability density curve has fatter tails and a higher peak at the mean than the normal distribution.

Limit cycle An attractor (for nonlinear dynamic systems) that has periodic cycles or orbits in phase space. An example is an undamped pendulum, which will have a closed-circle orbit equal to the amplitude of the pendulum's swing. See also *Attractor, Phase space.*

Lyapunov exponent A measure of the dynamics of an attractor. Each dimension has a Lyapunov exponent. A positive exponent measures sensitive dependence on initial conditions, or how much a forecast can diverge, based on different estimates of starting conditions. In another view, a Lyapunov exponent is the loss of predictive ability as one looks forward in time. Strange attractors are characterized by exhibiting at least one positive exponent. A negative exponent measures how points converge toward one another. Point attractors are characterized by all negative variables. See also *Attractor, Limit cycle, Point attractor, Strange attractor.*

Mechanical trading system A mechanical trading approach is one in which (i) a predetermined group of securities or markets are followed; (ii) mathematical formulas are applied to prices that then tell when to buy and when to sell; (iii) there are entry rules and exit rules for profitable and losing trades; (iv) there are rules for when to start trading and stop trading. The user of a mechanical trading system chooses a system and markets to trade and applies the system rules to market price action. If the system is computerized, data needs to be provided to the computer, to run the system, and to place the orders the system dictates. The key to creating a successful mechanical system is to avoid creating a curve-fitting system. A mechanical trading approach avoids the destructive emotionalism that permeates discretionary trading.

Membership function A function that represents the possibility of belonging to a fuzzy set. Membership functions are convex functions that range between zero and one.

Modern portfolio theory (MPT) The blanket name for the quantitative analysis of portfolios of risky assets based on the expected return (or mean expected value) and the risk (or standard deviation) of a portfolio of securities. According to MPT, investors would require a portfolio with the highest expected return for a given level of risk.

Moving average A trend-following indicator that works best in a trending environment. Moving averages smooth out price action but operate with a time lag. Any number of moving averages can be employed, with different time spans, to generate buy and sell signals. When only one average is employed, a buy signal is given when the price closes above the average. When two averages are employed, a buy signal is given when the shorter average crosses above the longer average. Technicians use three types: simple, weighted, and exponentially smoothed averages.

Mutation Mutation is an operator in genetic algorithms. It introduces diversity during reproduction. At a very low level of probability, binary mutation replaces bits on a chromosome with randomly-generated bits. This probability is a parameter of genetic algorithms.

Network Mathematically defined structure of a computing system where the operations are performed at specific locations (nodes) and the flow of information is represented by directed arcs.

Neural computing A fast-growing field of computing technology inspired by studies of the brain. The computing operations are performed by a large number of relatively simple, often adaptive, processing units. Neural computing, from its origins, is ideally adapted to doing pattern matching, pattern recognition, and control function synthesis.

Neural net trading system A neural net trading system is an automated way of trading a financial security or financial market based on neural network technology. A neural net trading system can be used as a decision-support or decision-making system.

Neural network A system that applies *neural computation*. An adaptive, nonlinear dynamical system. Its equilibrium states can recall or recognize a stored pattern or can solve a mathematical or computational problem.

Neuron A nerve cell in the physiological nervous system.

Nonlinearity Refers to a mapping which is nonlinear or in which the input is not a multiple of the output. A nonlinear network can be achieved by using nonlinear transfer functions, by competition among neurons, or by normalization.

Order parameter In a nonlinear dynamic system, a variable that summarizes the individual variables that can affect a system. For example, the Dow Jones is an order parameter, because it summarizes the change of some 30 stocks.

Oscillators Technical indicators that are utilized to determine when a market is in an overbought and oversold condition. Oscillators are plotted at the bottom of a price chart. When the oscillator reaches an upper extreme, the market is oversold. Two types of oscillators use momentum and rates of change.

Parallel processing   A form of computing in which many computations are being processed concurrently. One of the unique features of neural computing is that it provides an inherently clean and simple mechanism for dividing the computational task into subunits. This inherent parallelism makes it an ideal candidate for highly parallel architectures.

Pattern recognition The categorization of patterns in some domain into meaningful classes. A pattern usually has the form of a vector of measurement values.

Persistence The tendency of a series to follow trends. If the system has increased in the previous period, the chances are that it will continue to increase in the next period. Persistent time series have a long memory; long-term correlation exists between current events and future events. See also *Antipersistence, Hurst exponent, Rescaled range analysis.*

Phase space A graph that shows all possible states of a system. In phase space, the value of a variable is plotted against possible values of the other variables at the same time. If a system has three descriptive variables, the phase space is plotted in three dimensions, with each variable taking one dimension.

Point attractor In nonlinear dynamics, an attractor where all orbits in phase space are drawn to one point or value. Essentially, any system that tends to a stable, single-valued equilibrium will have a point attractor. A pendulum damped by friction will always stop. Its phase space will always be drawn to the point where velocity and position are equal to zero. See also *Attractor, Phase space.*

Price patterns Patterns that appear on price charts that have predictive value. Patterns are divided into reversal patterns and continuation patterns.

Processing element In neural computation: the neuron-like unit that, together with many other processing elements, forms a neural computing network. Computational abstraction of a neuron.

Random walk Brownian motion, where the previous change in the value of a variable is unrelated to future or past changes. See also *White noise.*

Rate of change A technique used to construct an overbought/oversold oscillator. Rate of change employs a price ratio over a selected span of time. To construct a ten-day rate of change oscillator, the last closing price is divided by the closing price ten days earlier. The resulting value is plotted above or below a value of 100.

Ratio analysis The use of a ratio to compare the relative strength between two entities. An individual stock or industry group divided by the S&P 500 index can determine whether that stock or industry group is outperforming or underperforming the stock market as a whole. Ratio analysis can be used to compare any two entities. A rising ratio indicates that the numerator in the ratio is outperforming the denominator. Ratio analysis can also be used to compare market sectors such as the bond market with the stock market or commodities with bonds. Technical analysis can be applied to the ratio line itself to determine important turning points.

Recall schedule A schedule which specifies how parameters associated with the response of a network change over the course of recall.

Relative-strength index (RSI) A popular oscillator developed by Welles Wilder, Jr., and described in his 1978 book, *New Concepts in Technical Trading Systems*. RSI is plotted on a vertical from 0 to 100. Values above 75 are considered to be overbought and values below oversold. When prices are over 75 or below 25 and diverge from price action, a warning is given of a possible trend reversal. RSI usually employs time spans of 9 or 14 days.

Reproduction A procedure applied in genetic algorithms whereby problem solutions or parents are selected from a population of problem solutions using a random selection procedure. Each parent's selection chances are biased so that the parents with the highest evaluations are most likely to reproduce. Children are made by copying the parents, and the parents are returned to the population.

Rescaled range (R/S) analysis The method developed by H.E. Hurst to determine long-memory effects and fractional Brownian motion. A measurement of how the distance covered by a particle increases over longer and longer time scales. For Brownian motion, the distance covered increases with the square root of time. A series that increases at a different rate is not random. See also *Antipersistence, Fractional Brownian motion, Hurst exponent, Persistence.*

Resistance The opposite of support. Resistance is marked by a previous price peak and provides enough of a barrier above the market to halt a price advance.

Roulette wheel parent selection A technique to determine which problem solutions or population members are chosen for reproduction. Using this technique, each chromosome's evaluation is proportional to the size of its slice on a roulette wheel. Selection of parents to reproduce is carried out through successive spins of the roulette wheel spinner. After a spin, the chromosome chosen is the one in whose slice the arrow ends up. See also *Genetic algorithm.*

Self-organization Adaptive change of structures in neural networks or their corresponding weight patterns in response to a learning stimulus. Usually this is unsupervised.

Set theory The study of sets or classes of objects. The set is the basic unit in math just as the symbol is the basic unit in logic. Logic and set theory make up the foundations of math. In theory all the symbols of advanced calculus and nuclear physics are just shorthand for the longhand of sets and logic. Classical set theory does not acknowledge the fuzzy or multivalued set whose members belong to the set to some degree. Classical set theory is bivalent. Each set contains members all or none.

Simple average A moving average that gives equal weight to each day's price data.

Simulated annealing A technique used to search for global minima in an energy surface in which states are updated based on a statistical rule rather than deterministically; this update rule changes to become more deterministic as the search progresses. During the optimization process, the objective function can be decreased or increased in order to find a global optimum.

Speech recognition Automatic decoding of a sound pattern into phonemes or words.

Stochastics An overbought/oversold oscillator that is based on the principle that as prices advance, the closing price moves to the upper end of its range. In a downtrend, closing prices usually appear near the bottom of their recent range. Time periods of 9 and 14 days are usually employed in its construction. Stochastics uses two lines: %K and its two-day average, %D. These two lines fluctuate in a vertical range between 0 and 100. Readings above 80 are overbought, while readings below 20 are oversold. When the faster %K crosses below the %D line and the lines are over 80, a sell signal is given. There are two stochastics versions: fast stochastics and slow stochastics. Most traders use the slower version because of its smoother look and more reliable signals.

Strange attractor An attractor in phase space, where the points never repeat themselves and the orbits never intersect, but both the points and the orbits stay within the same region of phase space. Unlike limit cycles or point attractors, strange attractors are nonperiodic and generally have a fractal dimension. They are a configuration of a nonlinear chaotic system. See also *Attractor, Chaos, Limit cycle, Point attractor.*

Subsethood The degree to which one set contains another set. In classical set theory a set has subsets all or none. In fuzzy logic it is a matter of degree. That means the subsethood or containment value can take any value between 0% and 100%. The measure of subsethood comes from the subsethood theorem. The subset theorem gives a new way to view the probability of an event. It equals the whole in the part. The probability of the part or event is the degree to which the whole or the "space" of all events is contained in the part. This relation cannot hold if subsethood is not fuzzy and can take on only the extreme black–white values of 0% and 100%.

Summation function The part of a processing element that adds the signals that enter the element.

Supervised learning Learning in which a system is trained by using a teacher to show the system the desired response to an input stimulus, usually in the form of a desired output.

Threshold A constant which is used as a comparison level by a variable. If the variable has a value above the threshold some action is taken (for example, a neuron fires), and if its value is below the threshold, no action is taken.

Trading system A trading system is an automated way of trading a financial security or financial market. Such a system can be designed on the basis of simple if–then rules, imbedded in a mechanical trading system or on the basis of neural networks, genetic algorithms, fuzzy logic or combinations of these technologies. A trading system can be used as a decision-support or decision-making system.

Training Exposing a neural computing system to a set of example stimuli to achieve a particular user-defined goal.

Transfer function The component of a processing element through which the sum is passed (transformed) to create net output. It is usually nonlinear.

Trend Refers to the direction of prices. Rising peaks and troughs constitute an uptrend; falling peaks and troughs constitute a downtrend. A trading range is characterized by horizontal peaks and troughs. Trends are generally classified into major (longer than 6 months), intermediate (1 to 6 months), or minor (less than a month).

Trendlines Straight lines drawn on a chart below reaction lows in an uptrend, or above rally peaks in a downtrend, that determine the steepness of the current trend. The breaking of a trendline usually signals a trend change.

Universe of discourse The range over which a variable is defined.

Unsupervised learning Learning in which no teacher is used to show the correct response to a given input stimulus; the system must organize itself purely on the basis of the input stimuli it receives. Often synonymous with clustering.

Weighted average A moving average that uses a selected time span but gives greater weight to more recent price data.

Weighted connections The channels through which information enters processing elements in a neural computing system, also called interconnects, throughout which memory is distributed.

White noise The audio equivalent of Brownian motion; sounds that are unrelated and sound like a hiss. The video equivalent of white noise is "snow" on a television receiver screen.

Bibliography

Adriaans P, Zantinge D (1996) Data mining, Syllogic, Addison-Wesley, Reading, MA, p 158

Altman EI (1968) Financial ratios, discriminant analysis and the prediction of corporate bankruptcy. Journal of Finance, September: 589–609

Amari S-I (1992) Neural theory of association and concept-formation. Biological Cybernetics 26: 175–185

Anderberg MR (1973) Cluster analysis for applications. Academic Press, New York

Andrews DF (1972) Plots of high-dimensional data. Biometrics, 28: 125–136

Ang JS, Patel KA (1975) Bond rating methods: Comparison and validation. Journal of Finance, May: 631–640

Araki S, et al. (1991) A self-generating method of fuzzy inference rules: fuzzy engineering toward human friendly systems. IFSE 2

Beder TS (1995) VAR: seductive but dangerous. Financial Analysts Journal 3: 12–24

Bernstein J (1992) Timing signals in the futures market. Probus, Chicago

Bernstein PL (1990) Flows of funds and flows of expectations. Journal of Portfolio Management 16

Bernstein P (1992) Capital ideas: the improbable origins of modern Wall Street. Free Press, New York

Berry M, Lindoff G (1997) Data mining techniques for marketing, sales and customer support. Wiley, Chichester, p 454

Bezdek JC (1981) Pattern recognition with fuzzy objective function. Algorithm, Plenum, New York

Bigus JP (1996) Data mining with neural networks. McGraw-Hill, New York, p 220

Bishop CM (1995) Neural networks for pattern recognition. Oxford University Press, Oxford, UK

Biswas G, Jain AK, Dubes RC (1981) Evaluation of projection algorithms. IEEE Transactions on Pattern Analysis and Machine Intelligence 3: 701–708

Blayo F, Demartines P (1991) Data analysis: how to compare Kohonen neural networks to other techniques. In Prieto (ed) Proceedings of IWANN'91, Lecture Notes in Computer Science, Springer-Verlag, Berlin, pp 469–476

Blayo F, Demartines P (1992) Algorithme de Kohonen: application à l'analyse de données économiques. Bulletin des Schweizerischen Elektrotechnischen Vereins & des Verbandes Schweizerischer Elektrizitatswerke 83(5): 23–26

Boyle PP (1977) Options: a Monte Carlo approach. Jour. Fin. Econ. 4: 323–338

Carlson E (1990) Neural networks in appraisal of forest woodlots (In Finnish). Hermoverkon käyttö metsäkiinteistöjen arvioinnissa. Maanmittaus 2/1990: 27–47

Carlson E (1991) Self-organizing feature maps for appraisal of land value of shore parcels. In: Kohonen T, Mäkisara K, Simula O, Kangas J, Artificial Neural Networks. North-Holland, Amsterdam, pp 1309–1312

Carlson E (1992) Neural thematic maps and neural GIS. Proceedings of the 4th Scandinavian Research Conference on GIS, Helsinki

Carlson E (1997) Scaling and sensitivity in appraisal. Proceedings of WSOM'97, Espoo, Finland

Chan KG, Karolyi GA, Longstaff FA, Sanders AB (1992) An empirical comparison of alternative models of the short-term interest rate. Journal of Finance 48: 1209–1227

Chang CL, Lee RCT (1973) A heuristic relaxation method for nonlinear mapping in cluster analysis. IEEE Transactions on Systems, Man, and Cybernetics 3: 197–200

Chatfield C (1985) The initial examination of data. Journal of the Royal Statistical Society 148, Part 3: 214–253

Chen H, Schuffels C, Orwig R (1996) Internet categorization and search: a self-organizing approach. Journal of Visual Communication and Image Representation 7: 88–102

Chernoff H (1973) The use of faces to represent points in k-dimensional space graphically. Journal of the American Statistical Association 68: 361–368

Cichocki A, Unbehauen R (1993) Neural networks for optimization and signal processing. Wiley, Chichester

Colin A (1992) Neural networks and genetic algorithms for exchange rate forecasting, Proceedings IJCNN, Beijing, China

Cottrell M, de Bodt E (1996) A Kohonen map representation to avoid misleading interpretations. In: Verleysen M (ed) Proc. ESANN'96, Editions D Facto, Bruxelles, pp 103–110

Cottrell M, de Bodt E (1997) A powerful tool for fitting and forecasting deterministic and stochastic processes: the Kohonen classification, submitted to International Congress on Artificial Neural Networks, Genève

Cottrell M, Ibbou S (1995) Multiple correspondence analysis of a crosstabulation matrix using the Kohonen algorithm. In: Verleysen M (ed) Proc. ESANN'95, Editions D Facto, Bruxelles, pp 27–32

Cottrell M, Letremy P (1994) Classification et analyse des correspondances au moyen de l'algorithme de Kohonen: application à l'étude de données socio-économiques. Proc. Neuro-Nîmes, pp 74–83

Cottrell M, Rousset P (1997) The Kohonen algorithm: a powerful tool for analysing and representing multidimensional quantitative and qualitative data. Proc. IWANN'97

Cottrell M, Letremy P, Roy E (1993) Analyzing a contingency table with Kohonen maps: a factorial correspondence analysis. In: Cabestany J, Mary J, Prieto A (eds) Proc. IWANN'93, Lecture Notes in Computer Science, Springer-Verlag, Berlin, pp 305–311

Cottrell M, Fort JC, Pagès G (1994) Two or three things that we know about the Kohonen algorithm. In: Verleysen M (ed) Proc. ESANN'94. Editions D Facto, Bruxelles, pp 235–244

Cottrell M, Girard B, Girard Y, Muller C, Rousset P (1995) Daily electrical power curves: classification and forecasting using a Kohonen map, from natural to artificial neural computation. In: Mira J, Sandoval F (eds) Proc. IWANN'95, Lecture Notes in Computer Science 930, Springer, Berlin, pp 1107–1113

Cottrell M, de Bodt E, Grégoire Ph (1996a) Simulating interest rate structure evolution on a long term horizon: a Kohonen map application. Proceedings of Neural Networks in the Capital Markets, Californian Institute of Technology, World Scientific Ed., Passadena

Cottrell M, de Bodt E, Henrion EF (1996b) Understanding the leasing decision with the help of a Kohonen map: an empirical study of the Belgian market. Proc. ICNN'96 International Conference, Vol.4, pp 2027–2032

Cottrell M, Fort JC, Pagès G (1996c) Two or three mathematical things about the Kohonen algorithm. Preprint SAMOS report # 55, Paris, 23pp

Cottrell M, Fort JC, Pagès G (1997) Theoretical aspects of the Kohonen algorithm. WSOM'97, Helsinki, pp 246–267

Cottrell GW, Munro P, Zipser D (1993) Image compression by back propagation: an example of extensional programming. In: Sharkey N (ed) Models of cognition: a review of cognitive science. Ablex, Norwood, NJ, vol. 1, pp 208–240

Cox E (1993) Adaptive fuzzy systems. IIEE Spectrum, February 1993

Cox JC, Ingersoll JE, Ross SA (1985a) An intertemporal general equilibrium model of asset prices. Econometrica 53: 363–384

Cox JC, Ingersoll JE, Ross SA (1985b) A theory of the term structure of interest rates. Econometrica 53: 385–407

Cumming S (1993) Neural networks for monitoring of engine condition data. Neural Computing & Applications 1(1): 96–102

Dahlquist M (1996) On alternative interest rate processes. Journal of Banking & Finance 20: 1033–1119

de Bodt E, Cottrell M (1997) A powerful tool for fitting and forecasting deterministic and stochastic processes: the Kohonen classification. Submitted to International Congress on Artificial Neural Networks, Genève

de Bodt E, Grégoire Ph, Cottrell M (1996) Simulating interest rate structure evolution on a long term horizon: a Kohonen map application. Proceedings of Neural Networks in the Capital Markets, Californian Institute of Technology, World Scientific Ed., Pasadena

Deboeck G (1991) Injecting new technology into a large organization: how many smoke signals does it take? Presented at Second Annual Conference on Expert Systems and Neural Networks in Trading, January 23-24, New York

Deboeck G (1993) Neural, genetic and fuzzy systems applications in Wall Street. PASE'93 Conference, Ascona, Switzerland. Neural Network World 3(6): 689-898

Deboeck G (1994a) Trading on the edge: neural, genetic and fuzzy systems for chaotic financial markets. Wiley, New York, p 377

Deboeck G (1994b) Investment Navigator, CD-ROM. Wyazata Technology, Grant Rapids, MN

Deboeck G (1995a) Advanced technologies for business reengineering. IBC's Sixth Annual Conference on Advanced Technologies for Trading & Asset Management, New York, September 14-15, and IBC's Annual Conference on Intelligent Systems in Business and Finance, London, February 13-14

Deboeck G (1995b) Business innovation at the World Bank, CD-ROM. World Bank

Deboeck G (1995c) Applicazione di technologie avanzate per la gestione della speculazione a breve termine del rischio e dei portafogli (Applications of advanced technology for forecasting, risk and portfolio management). Intelligenza Artificiale in Banca: Tendenze evolutive ed esperienze operative a comfronto, Universita Cattolica sel Sacro Cuore & Banque de france, Franco Angeli, Milano

Deboeck G (1996) Andere computers, anders communiceren, anders publiceren: IT-ontwikkelingen bij de Wereld Bank (New computers, new ways of communicating, different ways of publishing) Informatie: maandblad voor de informatievoorziening, vol. 38, Amsterdam, December, pp. 10-18

Deboeck G (1998) Financial Applications of Self-Organizing Maps, electronic newsletter. American Heuristics Inc, February, pp. 10-30

Deerwester S, Dumais ST, Furnas GW, Landauer TK, Harshman R (1990) Indexing by latent semantic analysis. Journal of the American Society for Information Science 41(6): 391-407

de Leeuw J, Heiser W (1982) Theory of multidimensional scaling. In: Krishnaiah PR, Kanal LN (eds) Handbook of statistics. North-Holland, Amsterdam, vol. 2, pp 285-316

Demartines P (1992) Organization measures and representations of Kohonen maps. In: Hérault JH (ed) First IFIP Working Group

Demartines P (1994) Analyse de données par réseaux de neurones auto-organisés (Data analysis through self-organized neural networks). PhD thesis, Institut National Polytechnique de Grenoble, Grenoble, France

Demartines P, Hérault J (1997) Curvilinear component analysis: a self-organizing neural network for nonlinear mapping of data sets. IEEE Transactions on Neural Networks 8: 148-154

DeMers D, Cottrell G (1993) Non-linear dimension reduction. In Hanson SJ, Cowan JD, Giles CL (eds) Advances in neural information processing systems 5: 580-587. Morgan Kaufmann, San Mateo, CA

Dixon JK (1979) Pattern recognition with partly missing data. IEEE Transactions on Systems, Man, and Cybernetics 9: 617-621

Doszkocs TE, Reggia J, Lin X (1990) Connectionist models and information retrieval. Annual Review of Information Science and Technology (ARIST) 25: 209-260

Duch W (1994) Quantitative measures for self-organising topographic maps. Open System and Information Dynamics 2 (3): 295-302

du Toit SHC, Steyn AGW, Stumpf RH (1986) Graphical exploratory data analysis. Springer-Verlag, New York

Dutta S, Shekhar S, Wong WY (1994) Decision support in non-conservative domains: generalization with neural networks. Decision Support Systems 11: 527-544

Everitt BS (1993) Cluster analysis. Edward Arnold, London

Fame E (1970) Efficient capital markets: a review of theory and empirical work. Journal of Finance, May, 383-417

Finch S, Chater N (1992) Unsupervised methods for finding linguistic categories. In: Aleksander I, Taylor J (eds) Artificial neural networks. North-Holland, Amsterdam, vol. 2, pp. 1365-1368

Flury B (1988) Common principal components and related multivariate models. Wiley Series in Probabilities and Mathematical Statistics, Wiley, Chichester

Fort J-C, Pagès G (1996) About the Kohonen algorithm: strong or weak self-organisation? Neural Networks 9(5): 773-785

Friedman JH (1987) Exploratory projection pursuit. Journal of the American Statistical Association 82: 249-266

Friedman JH, Tukey JW (1974) A projection pursuit algorithm for exploratory data analysis. IEEE Transactions on Computers 23: 881–890

Fyfe C, Baddeley R (1995) Non-linear data structure extraction using simple Hebbian networks. Biological Cybernetics 72: 533–541

Galbraith JK (1990) A short history of financial euphoria: financial genius is before the fall. Whittle Direct Books, Boston, MA

Gallant SI (1991) A practical approach for representing context and for performing word sense disambiguation using neural networks. Neural Computation 3: 293–309

Gallant SI, Caid WR, Carleton J, Hecht-Nielsen R, Pu Qing K, Sudbeck D (1992) HNC's MatchPlus system. ACM SIGIR Forum 26(2): 34–38

Gardes F, Gaubert P, Rousset P (1996) Cellulage de données d'enquîtes de consommation par une méthode neuronale. Preprint SAMOS report # 69

Garrido L, Gaitan V, Serra-Ricart M, Calbert X (1995) Use of multilayer feedforward neural nets as a display method for multidimensional distributions. International Journal of Neural Systems 6: 273–282

Gersho A (1979) Asymptotically optimal block quantization. IEEE Transactions on Information Theory 25: 373–380

Gia-Shuh Jang LF (1994) Intelligent trading of an emerging market. In: Deboeck G (ed) Trading on the edge: neural, genetic and fuzzy systems for chaotic financial markets. Wiley, Chichester, pp 80–101

Girardin L (1995) Mapping the virtual geography of the World-Wide Web. Proceedings of the Fifth International World Wide Web Conference, pp 131–139

Goodhill GJ, Sejnowski TJ (1996) Quantifying neighbourhood preservation in topographic mappings. In: Proceedings of the 3rd Joint Symposium on Neural Computation, University of California, pp. 61–82

Gray RM (1984) Vector quantization. IEEE ASSP Magazine, April, pp 4–29

Hafner K, Low M (1996) Where wizards stay up late: the origins of the Internet. Simon & Schuster, New York

Hansen LP (1982) Large sample properties of generalized method of moments estimator. Econometrica 50: 1029–1054

Hartigan J (1975) Clustering algorithms. Wiley, New York

Hastie T, Stuetzle W (1989) Principal curves. Journal of the American Statistical Association 84: 502–516

Haykin S (1994) Neural Networks: a comprehensive foundation. Macmillan, New York

Heath D, Jarrow R, Morton A (1992) Bond pricing and the term structure of interest rates: a new methodology for contingent claims valuation. Econometrica 60: 77–89

Hecht-Nielsen R (1989) Neurocomputing. Addison-Wesley, Reading, MA

Hecht-Nielsen R (1994) Context vectors: general purpose approximate meaning representations self-organized from raw data. In: Zurada JM, Marks II RJ, Robinson CJ (eds) Computational intelligence. IEEE Press, New York

Hecht-Nielsen R (1995) Replicator neural networks for universal optimal source coding. Science 269: 1860–1863

Ho TS, Lee S-B (1986) Term structure movements and pricing interest rate contigent claims. Journal of Finance 41: 1011–1029

Honkela T, Kaski S, Lagus K, Kohonen T (1996) Newsgroup exploration with WEBSOM method and browsing interface. Technical Report A32, Helsinki University of Technology, Laboratory of Computer and Information Science, Espoo, Finland

Hoptroff AR (1993) The principles and practice of time series forecastings and business modelling using neural nets. Neural Computing and Applications 1: 59–66

Hotelling H (1933) Analysis of a complex of statistical variables into principal components. Journal of Educational Psychology 24: 417–441, 498–520

IFC (1996a) Emerging stock market factbook 1996. International Finance Corporation

IFC (1996b) Emerging market database: the premier benchmark for global investors. IFC folder

Jain AK, Dubes RC (1988) Algorithms for clustering data. Prentice-Hall, Englewood Cliffs, NJ

Jang GS, Lai F (1993) Intelligent stock market prediction system using dual adaptive-structure neural networks. In: Proceedings of Second International Conference on Artificial Intelligence Applications on Wall Street, New York

Kandel A (1986) Fuzzy mathematical techniques with applications. Addison-Wesley, Reading, MA

Kaski S (1997) Data exploration using self-organizing maps. In: Acta Polytechnica Scandinavica, Mathematics, Computing and Management in Engineering Series No 82, Espoo, Finland, p 57

Kaski S, Lagus K (1996) Comparing self-organizing maps. In: Proceedings of ICANN'96, Lecture Notes in Computer Science, vol. 1112, pp 809–814. Springer, Berlin

Kaski S, Kohonen T (1995) Structures of welfare and poverty in the world discovered by the self-organizing map. Faculty of Information Technology, Helsinki University of Technology, Espoo, Finland

Kaski S, Honkela T, Lagus K, Kohonen T (1996) Creating an order in digital libraries with self-organizing maps. In: Proceedings of WCNN'96, World Congress on Neural Networks. Lawrence Erlbaum and INNS Press, Mahwah, NJ

Keppler M, Lechhner M (1997) Emerging markets: research, strategies and benchmarks. Irwin Professional Publishing, Chicago

Kimoto T, Asakawa K, Yoda M, Takeoka M (1990) Stock market prediction system with modular neural networks. In: Proceedings of the International Joint Conference on Neural Networks, Vol. 1. IEEE Network Council, San Diego, CA

Klimasauskas CC (1992a) Hybrid technologies: more power for the future. Advanced Technologies for Developers 1. High-Tech Communications, Sewickly, PA

Klimasauskas C (1992b) Accuracy and profit in trading systems. Advanced Technology for Developers 1. High-Tech Communications, Sewickly, PA

Klimasauskas C (1992c) Genetic function optimization for time series prediction. Advanced Technology for Developers, July. High-Tech Communications, Sewickly, PA

Klimasauskas C (1992d) Hybrid neuro-genetic approach to trading algorithms. Advanced Technology for Developers, November. High-Tech Communications, Sewickly, PA

Klimasauskas C (1992e) An excel macro for genetic optimization of a portfolio. Advanced Technology for Developers, December. High-Tech Communications, Sewickly, PA

Klimasauskas CC (1992f) Hybrid fuzzy encodings for improved backpropagation. Advanced Technology for Developers, September. High-Tech Communications, Sewickly, PA

Kohonen T (1989) Self organization and associative memory, 3rd edn. Springer Verlag, Berlin

Kohonen T (1990) The self organizing map. Proc. of the IEEE 78(9): 1464–1480

Kohonen T (1994) What generalizations of the self-organizing map make sense. In: Marinaro M, Morasso PG (eds) Proc. ICANN'94, Int. Conf. on Artificial Neural Networks. Springer, London, vol. I, pp 292–297

Kohonen T (1997) Self-organizing maps, Second edition. Springer, Berlin, p 362

Kohonen T, Hynninen J, Kangas J, Laaksonen J (1995) SOM_PAK: the self-organizing map program package. Helsinki University of Technology, Laboratory of Computer and Information Science. Available via anonymous ftp at internet address cochlea.hut.fi (130.233.168.48).

Kohonen T, Kaski S, Lagus K, Honkela T (1996) Very large two-level SOM for the browsing of newsgroups. In: von der Malsburg C, van Seelen W, Vorbrüggen JC, Sendhoff B (eds) Proceedings of ICANN'96, International Conference on Artificial Neural Networks. Springer, Berlin

Kohonen T, Erkki Oja, Olli Simula, Ari Visa, Jari Kangas (1997) Engineering applications of the self-organizing map. Manuscript submitted to a journal, Espoo

Kosko B (1992) Neural networks and fuzzy systems. Prentice-Hall, Englewood Cliffs, NJ

Kruskal JB (1964) Multidimensional scaling by optimizing goodness of fit to a nonmetric hypothesis. Psychometrika 29: 1–27

Kruskal JB, Wish M (1978) Multidimensional scaling. Sage University Paper series on Quantitative Applications in the Social Sciences, no. 07-011. Sage Publications, Newbury Park, CA

Lagus K, Honkela T, Kaski S, Kohonen T (1996) Self-organizing maps of document collections: a new approach to interactive exploration. In: Simoudis E, Han J, Fayyad U (eds) Proceedings of KDD'96, 2nd International Conference on Knowledge Discovery and Data Mining. AAAI Press, Menlo Park, CA

Lau C (ed) (1992) Neural networks: theoretical foundations and analysis. IEEE Press, New York

Lee RCT, Slagle JR, Blum H (1977) A triangulation method for the sequential mapping of points from N-space to two-space. IEEE Transactions on Computers 26: 288–292

Legendre P (1994) Applied Statistics 43: 23–257

Lehtonen T (1996) Market value of the vacation sites as an application of neurocomputing (In Finnish).

Lomakiinteistöjen arvo neuraalilaskennan sovelluksena. Masters thesis, Helsinki University of Technology

LIFE (1992) Fuzzy engineering toward human friendly systems. In: Proceedings of the 1st International Fuzzy Engineering Symposium, Yokohama, Japan, November 13–15, 1991, Proceedings of the 2nd International Conference on Fuzzy Logic and Neural Networks, Tizuka, Japan, July 17–22, 1992

Lin X, Soergel D, Marchionini G (1991) A self-organizing semantic map for information retrieval. In: Proceedings of the 14th Annual International ACM/SIGIR Conference on Research & Development in Information Retrieval, pp 262–269

Linde Y, Buzo A, Gray RM (1980) An algorithm for vector quantizer design. IEEE Transactions on Communications 28: 84–95

Linsker R (1988) Self-organization in a perceptual network. Computer, March, 105–117

Linsker R (1991) How to generate ordered maps by maximizing the mutual information between input and output. Neural Computation 1: 396–405

Litterman R, Scheinkman J (1988) Common factors affecting bond returns. Financial Strategies Group, Goldman Sachs, New York

Lloyd SP (1982) Least squares quantization in PCM. IEEE Transactions on Information Theory 28: 129–137

Lo AW (1991) Long-term memory in stock market price. Econometrica 59(5)

Longstaff FA, Schwartz ES (1992) Interest rates volatility and the term structure: a two-factor general equilibrium model. Journal of Finance 47: 1259–1282

MacQueen J (1967) Some methods for classification and analysis of multivariate observations. In: Le Cam LM, Neyman J (eds) Proceedings of the 5th Berkeley Symposium on Mathematical Statistics and Probability. Volume I: Statistics, pp 281–297. University of California Press, Berkeley, CA

Makhoul J, Roucos S, Gish H (1985) Vector quantization in speech coding. Proceedings of the IEEE 73: 1551–1588

Mao J, Jain AK (1995) Artificial neural networks for feature extraction and multivariate data projection. IEEE Transactions on Neural Networks 6: 296–317

Maren AJ, Harston CT, Pap RM (1990) Handbook of neural computing applications. Academic Press, San Diego, CA

Mar Molinero C, Apellániz GP, Serrano-Cinca C (1996) A multivariate analysis of Spanish bond ratings. OMEGA: International Journal of Management Science 24(4): 451–462

Martin del Brio B, Serrano-Cinca C (1993) Self-organizing neural networks for the analysis and representation of data: some financial cases. Neural Computing & Applications. Springer Verlag, Berlin, vol 1, pp 193–206

Martin del Brio B, Serrano-Cinca C (1995) Self-organizing neural networks: the financial state of Spanish companies. In: Refennes AP (ed) Neural networks in the capital markets. Wiley, Chichester

Merkl D (1993) Structuring software for reuse: the case of self-organizing maps. In: Proceedings of IJCNN'93 (Nagoya), International Joint Conference on Neural Networks, 3: 2468–2471. IEEE Service Center, Piscataway, NJ

Micklethwait J, Wooldridge A (1996) The witch doctors: What the management gurus are saying, why it matters and how to make sense of it. Mandarin Paperbacks, p 395

Moody J, Utans J (1995) Architecture selection strategies for neural networks: application to corporate bond rating prediction. In: Refennes AP (ed) Neural networks in the capital markets. Wiley, Chichester

Morningstar (1997) Ascent user's guide. Morningstar Inc, Chicago, IL

Mulier F, Cherkassky V (1996) Self-organization as an iterative kernel smoothing process. Neural Computation 7: 1165–1177

Nokaka I, Takeuchi H (1995) The knowledge creating company. Oxford University Press, Oxford, UK

Oja E (1983) Subspace methods of pattern recognition. Research Studies Press, Letchworth, UK

Oja E (1992) Principal components, minor components, and linear neural networks. Neural Networks 5: 927–935

Oja E (1994) Neural networks, principal components, and subspaces. International Journal of Neural Systems 1(1): 61–68

Pagès G (1993) Voronoï tesselation, space quantization algorithms and numerical integration. In: Verleysen M (ed) Proc. ESANN'93. Editions D Facto, Bruxelles, pp 221–228

Pardo R (1992) Design, testing and optimization of trading systems. Wiley, New York

Peters E (1991) Chaos and order in the capital markets: a new view of cycles, prices and market volatility. Wiley, New York

Peters EE (1994) Fractal market analysis. Wiley, New York

Pinches GE, Mingo KA (1973) A multivariate analysis of industrial bond ratings. Journal of Finance, March, pp 1–18

Pogue TF, Soldofsky RM (1969) What's in a bond rating? Journal of Financial and Quantitative Analysis, June, pp 201–228

Principe JC, Wang L (1995) Non-linear time series modeling with self-organization feature maps. In: Proc. NNSP'95, IEEE Workshop on Neural Networks for Signal Processing, pp 11–20. IEEE Service Center, Piscataway, NJ

Pruitt SW, White RF (1988) The CRISMA trading system: who says technical analysis can't beat the market? The Journal of Portfolio Management, Spring: 55–58

Prusak L (1997) Knowledge in organizations. Butterworth-Heinemann, Boston, MA

Rahimian E, Singh S, Thammachacote T, Virmani R (1993) Bankruptcy prediction by neural networks. In: Trippi R, Turban E (eds) Neural networks in finance and investing. Probus, Chicago

Refennes AP (ed) (1995) Neural networks in the capital markets. Wiley, New York

Refennes AN, Zapranis A, Francis G (1994) Stock performance modelling using neural networks: a comparative study with regression models. Neural Networks 7(2): 375–388

Regnier P (1996) The revised star rating. Morningstar Investor, November

Resta M (1997) Self-organizing evolutionary models in financial markets forecasting. Proceedings of WSOM'97, Workshop on Self-Organizing Maps, Helsinki University of Technology, pp 187–190

Ripley BD (1996) Pattern recognition and neural networks. Cambridge University Press, Cambridge, UK

Ritter H, Kohonen T (1989) Self-organizing semantic maps. Biological Cybernetics 61: 241–254

Ritter H, Martinetz T, Schulten K (1992) Neural computation and self-organizing maps: an introduction. Addison-Wesley, Reading, MA

Rubner J, Tavan P (1989) A self-organizing network for principal component analysis. Europhysics Letters 10: 693–698

Rumelhart DE, Hinton GE, Williams RJ (1986) Learning internal representations by error propagation. In: Rumelhart DE, McClelland JL, and the PDP Research Group (eds) Paralled distributed processing: explorations in the microstructure of cognition. Volume 1: Foundations, pp 318–362. MIT Press, Cambridge, MA

Salton G, McGill MJ (1983) Introduction to modern information retrieval. McGraw-Hill, New York

Samad T, Harp T (1992) Self-organization with partial data. Network: Computation in Neural Systems 3(2): 205–212

Sammon JW Jr (1969) A nonlinear mapping for data structure analysis. IEEE Transactions on Computers C-18(5): 401–409

Scholtes JC (1991) Unsupervised learning and the information retrieval problem. In: Proceedings of IJCNN'91, International Joint Conference on Neural Networks. IEEE Service Center, Piscataway, NJ

Scott DW (1992) Multivariate density estimation: theory, practice and visualisation. Wiley, New York

Serrano-Cinca C (1996) Self-organizing neural networks for financial diagnosis. Decision Support Systems 17: 227–238

Shepard RN (1962) The analysis of proximities: multidimensional scaling with an unknown distance function. Psychometrika 27: 125–140, 219–246

Singleton JC, Surkan AJ (1995) Bond rating with neural networks. In: Refennes AP (ed) Neural networks in the capital markets. Wiley, Chichester

Smith M (1993) Neural networks for statistical modeling. Van Nostrand Reinhold, New York

Sneath PHA, Sokal RR (1973) Numerical taxonomy. Freeman, San Francisco, CA

Standard and Poor (1995) Ratings y Análisis, No 5. Standard and Poor, Madrid

Sugeno M, et al. (1992) Fuzzy systems theory and its applications. Tokyo Institute of Technology

Templeton JM (1997) A grand old guru still favors emerging markets. New York Times, April 27

Torgerson WS (1952) Multidimensional scaling I: Theory and method. Psychometrika 17: 401–419

Trippi R, Turban E (eds) (1993) Neural networks in finance and investing: using artificial intelligence to improve real-world performance. Probus, Chicago

Tryon RC, Bailey DE (1973) Cluster analysis. McGraw-Hill, New York

Tukey JW (1977) Exploratory data analysis. Addison-Wesley, Reading, MA

Tulkki A (1996) Application of the self-organizing map for appraisal of dwellings (In Finnish). Itseorganisoivan kartan soveltuvuus asuntojen hintojen arviointiin. Masters thesis, Helsinki University of Technology

Ultsch A (1993a) Self-organizing neural networks for visualization and classification. In: Opitz O, Lausen B, Klar R (eds) Information and classification. Springer Verlag, Berlin, pp 307-313

Ultsch A (1993b) Self-organized feature maps for monitoring and knowledge aquisition of a chemical process. In: Gielen S, Kappen B (eds) Proceedings of the International Conference on Artificial Neural Networks. Springer, Berlin, pp. 864-867

Ultsch A, Siemon HP (1990) Kohonen's self-organizing feature maps for exploratory data analysis. In: Proceedings of INNC'90 Paris. Kluwer Academic, Dordrecht, pp 305-308

Utsugi A (1996) Topology selection for self-organising maps. Network: Computation in Neural Systems 7: 727-740

Utsugi A (1997) Hyperparameter selection for self-organising maps. Neural Computation 9(3): 623-635

Varfis A, Versino C (1991) Clustering of European regions on the basis of socio-economic data: a Kohonen feature map approach. In: Proceedings of the PASE International Workshop – Parallel Problem Solving from Nature: Applications in Statistics and Economics, pp 57-68

Varfis A, Versino C (1992) Clustering of socio-economic data with Kohonen maps. Neural Network World 2(6): 813-834

Vasiceck OA (1977) An equilibrium characterization of the term structure. Journal of Financial Economics 5: 177-188

Viitanen Kauko, Anttila Kari, Vuorio Kaisa (1994) Urban property market and land law in Finland. Department of Surveying, Helsinki University of Technology, Espoo, Finland

Wallace M (1996) Genetic algorithms, neural networks help traders on the edge. In: Emerging markets, Special IMF/World Bank Technology Supplement

Webb AR (1995) Multidimensional scaling by iterative majorization using radial basis functions. Pattern Recognition 28: 753-759

Welstead S (1993) Financial data modeling with genetically optimized fuzzy systems. In: Proceedings of the 2nd Annual International Conference on Artificial Intelligence Applications on Wall Street, New York, April 19-22

Wish M, Carroll JD (1982) Multidimensional scaling and its applications. In: Krishnaiah PR, Kanal LN (eds) Handbook of statistics, vol. 2, pp 317-345. North-Holland, Amsterdam

Wong F, Lee D (1993) A hybrid neural network for stock selection. In: Proceedings of the 2nd Annual International Conference on Artificial Intelligence Applications on Wall Street, New York, April 19-22

World Bank (1996) World Bank Development Report. Oxford University Press, Oxford, UK

Young FW (1985) Multidimensional scaling. In: Kotz S, Johnson NL, Read CB (eds) Encyclopedia of statistical sciences, vol. 5, pp 649-659. Wiley, New York

Young G, Householder AS (1938) Discussion of a set of points in terms of their mutual distances. Psychometrika 3: 19-22

Subject Index

A

Adaptive 233
Advertisements 143-144
Artificial Neural Networks (ANN) 12, 73, 107
Arbitrage opportunities 25, 113, 116
Ascent™ 40, 46
Asset allocation 101, 103
Attribute(s) xxxix, 45, 121
Audio-visual computing xxviii
Automatic voice recognition 10

B

BACH data base 16
Banking system 72, 83, 95
Bankruptcy xxix, 4, 19, 23
Batch training 167, 185
Bear market decile 56
Benchmark 39, 44, 52, 71, 105
Best practices 203
Bond rating 3, 4, 11, 12, 13
Bonds international 12, 49
Bookmark 170
Box-and-whiskers plot 4
Brand-name recognition 143, 152, 156
Browsing 168, 169, 170

C

Catalogues 168
Chaos xxvi, 233
Chinese consumer market 141
Chinese-European Business School (CEIBS) 143, 156
Classifier 116
Clustering xxviii, xxix, xxxi, xli, 4, 7, 28, 78, 103, 144, 145, 179, 188, 203
  significance 148
  -size 209
Code book vectors xxxii
Collaboration xxv, xxvi, xxvii

Color coding
  - of SOM maps 188, 195, 197, 203
  - color centers 197-198
Commerce on WWW xxv
Communication links 161, 162
Companies 9
  - large, moderate-size, small 100
Compatibility 35
Competitive advantage 106
Competitive learning xxxiv, 26, 40, 234
Competitive neural networks xxxvi
Complex data set 159
Component plane representation 54, 131, 132, 149, 183, 222
Computational efficiency 181
Connector words 176
Construction 83, 95, 97
Consumer xxix, 142
  - business dinners 154
  - dining habits 154, 156
  - lifestyle 154
  - preferences xli
  - private dinner 154
  - product attributes 142, 152
  - segmentation 141-142
  - scoring 189
  - survey 141
Conventional methods 159
Convergence 3, 14, 18, 109
Corporate
  - failure 59, 71
  - rating 71
  - strategy 3
Correlation coefficient 54, 85, 131
Correlating variable 128
Country credit risk analysis 203, 212, 218
Cox-Ingersoll-Ross interest rate model 38
Credit risk 11, 203
Currency risk 85
Curvature 30, 223
Customer (see consumer)
Cybershops xxv

Author Index

Website Index (as of April 1998)

Artificial intelligence:

for beginners:
http://home.clara.net/mll/ai/

Data Mining Tools and Vendors:

BusinessMiner (Business Objects, Inc):
http://www.businessobjects.com

Clementine (Integral Solutions Ltd.):
http://www.isl.co.uk

Darwin(Think Machines Corporation):
http://www.think.com

DataBase Mining Workstation (HNC
Software Inc.):
http://www.hnc.com

DataMind, DataCruncher (DataMind
Corporation):
http://www.datamindcorp.com

Data Mine, Data Mine Builder (Red Brick
Systems Inc):
http://www.RedBrick.com

IDIS (Information Discovery Inc.):
http://www.datamining.com

Intelligent Miner (IBM Corporation):
http://www.ibm.com

KnowledgeMiner:
http://www.scriptsoftware.com/km/index.html

KnowledgeSeeker (Angoss Software
International Ltd.):
http://www.angoss.com

MineSet (Silicon Graphics Computer Systems):
http://www.sgi.com

NeuralWorks Predict, Professional II+,
Explorer (NeuralWare Inc.):
http://www.neuralware.com/

SAS, JMP (SAS Institute Inc.):
http://www.sas.com/offices/intro.html

SPSS CHAID (SPSS Inc.):
http://www.spss.com

Trajan Software Ltd.:
http://www.trajan-software.demon.co.uk

Neural Networks:

Introduction to neural networks:
http://www.cs.stir.ac.uk/~lss/NNIntro/
InvSlides.html

Artificial neural networks technology:
DACS Technical Report Summary:
http://www.mcs.com/~drt/bprefs.html

FAQ on neural networks:
ftp://ftp.sas.com/pub/neural/FAQ.html
http://www.utica.kaman.com/techs/neural/
neural.html

Neural network societies and organizations:
http://www.emsl.pnl.gov:2080/docs/cie/
neural/societies.html

IEEE Neural Network Council:
http://www.ieee.org/nnc/

Neural Networks Research on the IEEE
Neural Network Council:
http://www.ieee.org/nnc

World-wide lists of groups doing research
on neural networks, Foundation for Neural
Networks (SNN):
http://www.mbfys.kun.nl/snn/pointers/
groups.html

Genetic Algorithms:

http://www.cs.gmu.edu/research/gag
http://www.shef.ac.uk/~gaipp/galinks.html
http://www.cs.bgu.ac.il/~omri/NNUGA/
http://divcom.otago.ac.nz:800/COM/
INFOSCI/SMRL/people/andrew/
publications/faq/hybrid/hybrid.htm

Self-Organizing Map:

Articles

Comp.ai.neural-nets:
http://websom.hut.fi/websom/
comp.ai.neural-nets-new/html/root.html

Applications

World poverty map:
http://www.cis.hut.fi/nnrc/worldmap.html

Tools

Bayesian SOM (Akio Utsugi) :
http://www.aist.go.jp/NIBH/~b60616/Lab/
index-e.html

Neural Networks Research Center, Helsinki
University of Technology:
http://nucleus.hut.fi/nnrc/
http://www.cis.hut.fi/nnrc/som.html

NeuroNet – Software:
http://www.neuronet.ph.kcl.ac.uk/neuronet/
software/software.html

NeuroShell:
http://www.wardsystems.com

NeuroSolutions:
http://www.nd.com/demo/demo.htm

SOM Toolbox:
http://www.cis.hut.fi/projects/somtoolbox/
Viscovery SOMine (Eudaptics Software
Gmbh):
http://www.eudaptics.co.at

WEBSOM:
http://nodulus.hut.fi/websom/

Statistical Software:

StatLib repository at Carnegie Mellon
University:
http://lib.stat.cmu.edu

SPSS Inc.:
http://www.spss.com

StatSoft Inc., on-line Electronic Statistics
Textbook: http://www.statsoft.com/textbook/
stathome.html

Clustering : R- Package:
http://alize.ere.umontreal.ca/~casgrain/en/
R/v4/modules.html

Societies:

Classification Society of North America:
http://www.pitt.edu/~csna/

Neural Information Processing Society
http://www.cs.cmu.edu/afs/cs/project/cnbc/
nips/nips-foundation.html